Bioengineering Case Studies

Wendi Goldsmith • Donald Gray • John McCullah

Bioengineering Case Studies

Sustainable Stream Bank and Slope Stabilization

Springer

Wendi Goldsmith
Bioengineering Group
Salem, MA, USA

John McCullah
Salix Applied Earthcare
Redding, CA, USA

Donald Gray
Department of Civil and Environmental Engineering
The University of Michigan
Ann Arbor, MI, USA

ISBN 978-1-4614-7995-6 ISBN 978-1-4614-7996-3 (eBook)
DOI 10.1007/978-1-4614-7996-3
Springer New York Heidelberg Dordrecht London

Library of Congress Control Number: 2013949146

Printed on acid-free paper

Springer is part of Springer Science+Business Media (www.springer.com)

Contents

TABLE 1

BIOENGINEERING CASE STUDIES: BY GEOGRAPHIC REGION AND COMPLETION DATE

GREAT LAKES AREA:

PROJECT NUMBER	PROJECT NAME	TYPE OF PROJECT	TECHNIQUES EMPLOYED	GEOGRAPHIC LOCATION	COMPLETION DATE
1	Fleming Creek	Stream bank repair and protection	***Rock vanes Pole planting***	Northfield TWP, MI	May 2006
2	Gateway Garden	Runoff and erosion control	***Stepped pools Riparian planting***	Ann Arbor, MI	Aug 2003
3	School Girls Glen	Erosion control and slope failure repair	***Boulder cascade VMSE Live staking Revegetation Invasive removal***	Ann Arbor, MI	Sept 2002
4	River Landing	Riverbank restoration and stabilization	***Vegetated riprap Live fascine ECB Revegetation***	Ann Arbor, MI	Oct 2005
5	Nichols Drive	Stream bank and road stabilization	***Rock vanes VMSE Chimney drain Revegetation***	Ann Arbor, MI	May 2008
6	Harvard Road	Gully repair and watershed restoration	***Rocked pools Sinuous channel ECB Live staking***	Ann Arbor, MI	Oct 2009
7	Malletts Creek	Stream bank repair and stabilization	***Rock vanes Low flow channel Coir logs ECB w/live stakes***	Malletts Creek watershed Southeastern Ann Arbor, MI	2011
8	Toboggan Hill	Erosion control	***Live fascines Live stakes***	Garden City, MI	2008
9	Arbor Cascades	Dam bypass channel	***Stepped pools Boulder spurs ECBs and grass seeding***	Ann Arbor, MI	2011

Western USA

PROJECT NUMBER	PROJECT NAME	TYPE OF PROJECT	TECHNIQUES EMPLOYED	GEOGRAPHIC LOCATION	COMPLETION DATE
10	Asaayi Lake	Watershed repair and restoration	***Landform grading Rock ramps Revegetation***	Chuska Mtns. New Mexico	2006
11	Hollywood Hills	Mass grading and canyon fills	***Landform grading Drainage swales Revegetation***	Lake Hollywood Los Angeles, CA	2005
12	Geyserville	Stream bank stabilization	***Rock vanes LSTP Live siltation Pole planting***	Geyserville, CA	2010
13	Buckhorn Mountain	Hwy. cut slope repair and stabilization	***Modified VMSE ("soil flaps") Veg. gabion baskets Soil amendments***	Hyw. 299W 0.1 mi below summit, Shasta Co, CA	Aug 2002
14	Buckhorn Summit	Hwy. cut slope repair and stabilization	***Live staking Brushlayers Soil amendments Stepped slope***	Hyw. 299W at summit Shasta Co, CA	2005
15	Stafford Watershed	Landslide repair and slope stabilization	***Pole drains Brushlayers Live staking***	Humbolt County, CA	2004
16	Pacifica	Landslide repair and slope stabilization	***Slope drains Live fascines Live staking***	San Mateo County, CA	Sept 1997
17	Branciforte Creek	Stream bank stabilization	***Veg. riprap w/pole planting and brushlayering Veg. gabion baskets***	Trib. to San Lorenzo R. in San Lorenzo Watershed	2002
18	San Vicente Creek	Stream bank stabilization	***Rock Vanes Rootwad revet. Veg. riprap - bent pole method Live staking***	Second-order stream in Santa Cruz Co nr city of Davenport, CA	2003

Western USA (cont)

19	Opal Cliffs	Slope stabilization	***Anchored TRM Live staking Hydroseeding***	Opal Cliffs Dr. Santa Cruz, CA	Aug 2007
20	Sulfur Creek	Stream bank stabilization	***Pole planting LWD and rock strucs Newbury rock riffle LPSTP w/live silt Stream realignment***	Turtle Bay, Redding Arboretum Redding, CA	2001
21	Secret Canyon	Slope repair and stabilization	***Toe protect w/brush layers Woven willow fences Pole planting***	Tributary to Sulphur Creek Redding, CA	2002

Eastern USA

22	Greenfield Road	Hwy. cut slope repair and stabilization	***Live brushlayers Live fascines Coir netting***	Colrain, MA	Dec 1989
23	Buffalo Bayou	Riverbank repair and stabilization	***Vegetated MSE Chimney drain Live fascines Live staking Live boom***	Buffalo Bayou Houston, TX	1996
24	Little Topashaw	Riverbank repair and stabilization	***LW structures Willow posts and poles***	Topashaw Cr. Yalobusha R. watershed central, MS	2006
25	New Concord	Hwy embankment stabilization	***Willow poles***	Intersection I-70 & SR83 near New Concord, OH	2003
26	Water Treatment Facility	Shoreline stabilization	***Coconut fiber rolls Bioswales Green roof Live brushlayering***	New Haven, CT	1999

Eastern USA (cont)

PROJECT NUMBER	PROJECT NAME	TYPE OF PROJECT	TECHNIQUES EMPLOYED	GEOGRAPHIC LOCATION	COMPLETION DATE
27	Walden Pond	Shoreline stabilization	***Vegetated MSE Live staking Live brushlayering Blueberry sod***	Concord, MA	1993
28	Hearthstone Quarry Brook	Channel stabilization	***Cross vanes Coconut fiber rolls Veg. gabion baskets Live brushlayering***	Chicopee, MA	1993
29	Mill Creek	Stream bank stabilization	***Vegetated MSE ECBs Live poles Live fascines and stakes***	Cincinnati, OH	1996
30	Charles River	Riverbank stabilization	***Coconut fiber rolls Live brushlayering Live staking ECBs***	Watertown, MA	1998
31	Connecticut River	Stream bank stabilization Erosion control	***Vegetated MSE Live brushlayering Live staking***	Northfield, MA	1995
32	Cumberland River	Stream bank stabilization	***Willow poles Live staking Live brushlayering Live fascines***	Nashville, TN	1996
33	Manhan River	Stream bank stabilization	***Coconut fiber rolls Bendway weirs Rock vanes***	Easthampton, MA	1997
34	Walgreen Slope	Erosion control	***Live fascines Live poledrains Live staking ECBs***	Marshfield, MA	2007
35	Creek Road	Stream bank repair and stabilization	***Rock vanes Bendway weirs Revegetation LSTP***	Yorkshire Cattaraugus Co New York	2009

TABLE 2

BIOENGINEERING CASE STUDIES: BY PROJECT DESIGNER AND/OR DIRECTOR

GREAT LAKES AREA:

PROJECT NUMBER	PROJECT NAME	TYPE OF PROJECT	GEOGRAPHIC LOCATION	CLIENT/ OWNER	PROJECT DESIGNER AND/OR DIRECTOR
1	Fleming Creek	Stream bank stabilization	Northfield TWP, MI	University of Michigan	*Sarah Weiss*, School of Natl. Resources University of Michigan and *Donald Gray*, Dept. of Civil and Envl. Engr. University of Michigan
2	Gateway Garden	Runoff and erosion control	Ann Arbor, MI	University of Michigan	Ayres, Lewis, Norris & May Inc. Construction & Bldg. Engineers Ann Arbor, MI
3	School Girls Glen	Erosion control and slope failure repair	Ann Arbor, MI	University of Michigan	*Tammy Orlow*, School of Natural Resources University of Michigan and *Donald Gray*, Dept. of Civil and Envl. Engr. University of Michigan
4	River Landing	Riverbank restoration and stabilization	Ann Arbor, MI	University of Michigan	Envl. Consulting & Technology [ECT] Environmental Consultants Ann Arbor, MI
5	Nichols Drive	Stream bank and road stabilization	Ann Arbor, MI	University of Michigan and City of Ann Arbor	Envl. Consulting & Technology [ECT] Environmental Consultants Ann Arbor, MI
6	Harvard Road	Gully repair and watershed restoration	Ann Arbor, MI	University of Michigan and City of Ann Arbor	OHM Advisors Architects, Engineers & Planners Ann Arbor, MI
7	Toboggan Hill	Erosion Control	Garden City, MI	Wayne County Parks Dept.	NTH Consultants Infrastructure Engr. & Envl. Services Northville, MI
8	Malletts Creek	Stream bank repair and stabilization	Malletts Creek watershed Southeastern Ann Arbor, MI	City of Ann Arbor	OHM Advisors Architects, Engineers & Planners Ann Arbor, MI
9	Argo Cascades	Dam bypass channel	Ann Arbor, MI	City of Ann Arbor	Recreation Engineering and Planning Stream Design & Restoration Boulder, CO

Western USA

PROJECT NUMBER	PROJECT NAME	TYPE OF PROJECT	GEOGRAPHIC LOCATION	CLIENT/ OWNER	PROJECT DESIGNER AND/OR DIRECTOR
10	Asaayi Lake	Watershed repair and restoration	Chuska Mtns. New Mexico	Navajo Nation	*H.J. Schor* H.J. Schor Consulting Anaheim, CA
11	Hollywood Hills	Mass grading and canyon fills	Lake Hollywood Los Angeles, CA	Los Angeles DWP	*Horst J. Schor* H.J. Schor Consulting Anaheim, CA
12	Geyserville	Stream bank stabilization	Geyserville, CA	CALTRANS	*John McCullah* Salix Applied Earthcare Redding, CA
13	Buckhorn Mountain	Hwy. cut slope repair and stabilization	Hyw. 299W 0.1 mi below summit, Shasta Co, CA	CALTRANS	*John McCullah* Salix Applied Earthcare Redding, CA
14	Buckhorn Summit	Hwy. cut slope repair and stabilization	Hyw. 299W at summit Shasta Co, CA	CALTRANS	*John McCullah* Salix Applied Earthcare Redding, CA
15	Stafford Watershed	Landslide repair and slope stabilization	Humbolt County, CA	Pacific Lumber Company	*John McCullah* Salix Applied Earthcare Redding, CA
16	Pacifica	Landslide repair and slope stabilization	San Mateo County, CA	Local homeowners association	Alan Kropp Alan Kropp & Assoc [AKA] Berkeley, CA
17	Branciforte Creek	Stream bank stabilization	Trib. to San Lorenzo R. in San Lorenzo watershed	Public Works Department Santa Cruz County	*John McCullah* Salix Applied Earthcare Redding, CA
18	San Vicente Creek	Stream bank stabilization	Second-order stream in Santa Cruz Co nr city of Davenport, CA	Public Works Department Santa Cruz County	*John McCullah* Salix Applied Earthcare Redding, CA
19	Opal Cliffs	Slope stabilization	Opal Cliffs Dr. Santa Cruz California	Local homeowner	*John McCullah* Salix Applied Earthcare Redding, CA
20	Sulfur Creek	Stream bank stabilization	Turlle Bay Redding, CA	Sacramento Watershed Action Group (SWAG)	*John McCullah* Salix Applied Earthcare Redding, CA
21	Secret Canyon	Slope repair and stabilization	Tributary to Sulphur Creek Redding, CA	CRMP and SWAG	*John McCullah* Salix Applied Earthcare Redding, CA

Eastern USA

22	Greenfield Road	Hwy. cut slope repair and stabilization	Colrain, MA	Massachusetts State Department of Highways	Robbin B. Sotir & Assoc. Soil Bioengineering Consultants Marietta, GA
23	Buffalo Bayou	Riverbank repair and stabilization	Buffalo Bayou Houston, TX	Riparian home owner	Robbin B. Sotir & Assoc. Soil Bioengineering Consultants Marietta, GA
24	Little Topashaw	Riverbank repair and stabilization	Topashaw Cr. Yalobusha R. watershed, no central, MS	USDA-NRCS and Chickasaw Co. Conservation District	F. Douglas Shields, Jr. Res. Hydraulic Engineer USDA-ARS Natl. Sedimentation Lab Oxford, Miss
25	New Concord	Hwy embankment stabilization	Intersection I-70 & SR83 near New Concord, OH	Ohio DOT	*Patrick Fox, T.H. Wu* Dept. of Civil & Envl. Engr. Ohio State University Columbus, OH
26	Water Purification Facility	Shoreline stabilization	New Haven, CT	South Central Connecticut Regional Water Authority	*Wendi Goldsmith* Bioengineering Group Salem, MA
27	Walden Pond	Shoreline restoration	Concord, MA	Massachusetts Department of Conservation and Recreation	*Wendi Goldsmith* Bioengineering Group Salem, MA
28	Hearthstone Quarry Brook	Channel stabilization	Chicopee, MA	Department of Public Works City of Chickopee	*Wendi Goldsmith* Bioengineering Group Salem, MA
29	Mill Creek Landfill	Stream bank stabilization	Lower Mill Creek Cincinnati, OH	Municipal Sewer District City of Cincinnati	*Wendi Goldsmith* Bioengineering Group Salem, MA
30	Charles R. Watertown Arsenal	Riverbank stabilization	Watertown, MA	US Army Corps of Engineers New England District	*Wendi Goldsmith* Bioengineering Group Salem, MA
31	Connecticut River	Stream bank stabilization	Northfield, MA	Western Mass. Electric Co.	*Wendi Goldsmith* Bioengineering Group Salem, MA

Eastern USA (cont)

PROJECT NUMBER	PROJECT NAME	TYPE OF PROJECT	GEOGRAPHIC LOCATION	CLIENT/ OWNER	PROJECT DESIGNER AND/OR DIRECTOR
32	Cumberland R. Shelby Bottoms Park	Bank stabilization	Nashville, TN	US Army Corps of Engineers Nashville District	*Wendi Goldsmith* Bioengineering Group Salem, MA
33	Manhan River	Bank stabilization	Easthampton, MA	Tennessee Gas Pipeline Company	*Wendi Goldsmith* Bioengineering Group Salem, MA
34	Walgreens Slope	Erosion control	Marshfield, MA	BRT Management	*Wendi Goldsmith* Bioengineering Group Salem, MA
35	Creek Road	Stream bank repair and stabilization	Yorkshire Cattaraugus Co New York	Town of Yorkshire, NY	*Todd Swackhamer* McMahon & Mann Consulting Engineers Buffalo, NY

Preface

Vegetation is likely the most natural, self-sustaining, and cost-effective method for protecting exposed soils, slopes and, streambanks from the erosive forces of raindrops, the tractive forces of flowing water, and the exfiltration of subsurface water.

Vegetative measures for stabilization purposes have developed rapidly during the past three decades, and have been applied frequently in practice to stabilize stream banks and restore watersheds so too have the use of related measures to stabilize upland slopes. Several guidance manuals and "how to do it" books have been published during this time period. However, to date no detailed compilation of the results of projects has been published. **Bioengineering Case Studies: Sustainable Stream Bank and Slope Protection** addresses this need. The book describes many different types of case histories of biostabilization projects and their outcomes including the restoration and maintenance of geomorphic and ecological functions.

In this book two experienced practitioners along with the principal author of two prior books on biotechnical engineering have collaborated to share a diverse set of project case studies from several regions of the USA. These projects were developed for different purposes and by project owners with distinct needs, constraints, and preferences. The projects described in this book were selected in order to reflect not only sound application of typical treatments but also to convey some more nuanced variations and innovations that reflect the practice of bioengineering in the twenty-first century.

The main purpose of **Bioengineering Case Studies: Sustainable Stream Bank and Slope Protection** is twofold, namely, to build upon the wealth of information provided in prior guidance manuals (such as the 2005 report 544, Environmentally Sensitive Channel and Bank Protection Measures, by the National Cooperative Highway Research Program (NCHRP)), and to describe different types of projects using functional living vegetation and evaluate their performance. The original NCHRP 544 report provided a detailed literature review of available knowledge and methods relating to environmentally sensitive channel and stream bank protection practices and their use by different practitioners and agencies. The NCHRP report mainly targets highway design engineering criteria. This book, on the other hand, focuses more on short- and long-term performance history, maintenance issues, and lessons learned from actual projects.

This book is a retrospective compilation of case studies. It is not intended to provide design procedural descriptions nor does it attempt to train practitioners on the many disciplines of science, engineering, construction, horticulture, and regulatory compliance that underlie each project. The objective is to help people visualize how projects have been configured in the past, and to understand precedents and examples so the approach will come across as both attractive and manageable. There are many guidance manuals for the reader to refer to; several are listed in the appendices or in the references at the end of each case study.

We would like to acknowledge the assistance of the following persons who helped directly in the preparation and review of the book. Horst Schor, HJ Schor Consulting, Anaheim, CA reviewed the write-ups of the Asaayi Lake (#10) and Hollywood Hills (#11) projects that he also designed. Alan Kropp, Alan Kropp & Assoc., Berkeley, CA provided useful information about the Pacifica (#16) project. Dr. F.D. Shields, Jr., USDA-ARS Natl. Sedimentation Lab, Oxford, MS provided helpful information and photo documentation about the Little Topashaw (#24) project. Prof. T.H. Wu, Dept. Civil Engr., Ohio State University, Columbus, OH provided useful information about the New Concord (#25) project that he directed. Molly Robinson, Manager, Water Treatment Services, City of Ann Arbor, reviewed the write-up and provided helpful information about the Argo Cascades (#9) project. Todd Swackhamer, Project Manager, MacMahon and Mann, Consulting Engineers, provided a write-up for the Creek Road (#35) project. Duke Bitsko, Corey Schutzman, Danielle Kelleher, and Amanda Bibbins from Bioengineering Group all provided valuable support hunting for project information from archives, selecting useful photographs, and editing text to shape multiple project documents into succinct case studies. The staff of Salix Applied Earthcare provided much useful information for the case studies. They sometimes worked as Project Managers for the Sacramento Watersheds Action Group (SWAG). This nonprofit watershed restoration group restored over 2-miles of Sulphur Creek (Proj #20) by soliciting grant funding, acquiring environmental permits, implementing the projects, and then monitoring effectiveness.

Redding, CA	John McCullah
Ann Arbor, MI	Donald Gray
Salem, MA	Wendi Goldsmith

About the Authors

Wendi Goldsmith is founder and CEO of Bioengineering Group, a Salem, Massachusetts based consulting firm whose mission is "Building sustainable communities on an ecological foundation." Since becoming involved in bioengineering practice in 1987, she has often led interdisciplinary collaborative design teams to adopt ecologically rich strategies. Her roles span planning of large scale multi-state watershed management and restoration projects, to design of stabilization and ecological enhancement treatments for riparian sites compromised by solid and hazardous waste, to leading infrastructure design procedures that account for future hydrology and sea level rise considerations. She is known for building consensus among diverse and often antagonistic stakeholder groups to help advance some of the largest public infrastructure projects in history. Ms. Goldsmith provided technical leadership for the first large scale federal project to provide resilient and adaptable infrastructure featuring hard structures and also extensive coastal wetland restoration and management, the Greater New Orleans Hurricane Storm Damage Risk Reduction System, featuring $14 billion of construction.

Ms. Goldsmith's academic training and applied experience have honed her expertise in the physical, chemical, and biological processes that affect watershed quality, including restoration techniques. She has extensive experience in all phases of project design and implementation for ecological restoration, particularly in dense urban environments, as well as for stabilization of lakes, rivers, and coastal areas. She has played a key role in promoting local familiarity and acceptance of bioengineering methods and for systems-scale sustainable solutions. Evaluating change in land use and its effect on geomorphic stability, nonpoint source pollution, and habitat degradation has been an integral part of Ms. Goldsmith's waterways assessments, stormwater management, restoration, and climate change adaptation projects. She is skilled in the areas of soil science, fluvial geomorphology, biogeochemistry, landscape design, wetland management, construction, and economics. She also has a thorough understanding of federal, state, and local environmental regulatory policies, and she has provided expert testimony on their meaning and intent on cases connected to damaged ecosystems. She has authored many peer-reviewed articles, book chapters, and lay articles; lectured at universities; and led training programs for professional audiences around the world. Her projects have received numerous awards including the AIA Top Ten Green Award, the American Academy of Environmental Engineering Award, and the American Consulting Engineers Council Grand Conceptor Award, among others.

Ms. Goldsmith holds an MS degree in plant and soil science from UMass, an MA degree in landscape design from the Conway School, and a BA degree in geology and geophysics/studies in the environment from Yale University. Additionally she trained under Europe's leading river and coastal bioengineering expert, Lothar Bestmann. She is a licensed Professional Geologist and holds multiple professional certifications.

Dr. Donald Gray is a Professor (Emeritus) of Civil and Environmental Engineering at the University of Michigan. He specializes in geotechnical engineering with a focus on slope stability, soil erosion, and bioengineering. Dr. Gray has taught graduate and undergraduate courses on the engineering properties of soils, soil/site improvement, slope stability, engineering geology, foundation engineering, and bioengineering for slope protection at the University of Michigan as well as many professional short courses and workshops. He has carried out research for decades on soil reinforcement and the influence of woody vegetation on the stability of slopes. His principle research interests lie in the areas of slope stability, earth reinforcement, and bioengineering.

Dr. Gray has served as a consultant to the National Park Service, US Forest Service, Natural Resources Conservation Service, US Army Corps of Engineers, and most recently to the California Levee Vegetation Research Program. Professor Gray's industrial experience includes a position as Petroleum Engineer with Mobil International Oil Company. He was also employed as Research Engineer with Chevron Research Company. He has served as a consultant on numerous occasions to both private and government agencies. On many occasions he provided expert testimony related to a broad range of engineering and resource management topics.

Dr. Gray is the principal author of two leading reference books on slope biostabilization which helped establish a basis for widespread adoption of design and construction methods in the US. More recently he was co-author of a publication entitled *Landforming: An Environmental Approach to Hillside Development, Mine Reclamation, and Watershed Restoration*. He was also a principal investigator in a 3-year project funded by the National Cooperative Highway Research Program on "Environmentally Sensitive Channel and Streambank Protection Measures." The goal of this project was to produce a guidance manual on this topic for use by highway engineers and their consultants.

He received all his degrees, including a BS in Geological Engineering and a PhD in Civil Engineering, from the University of California at Berkeley. His extensive and pioneering research combined with his well regarded publications has established him as the pre-eminent authority on bioengineering of our time.

John McCullah is President of Salix Applied Earthcare, a consulting firm located in Redding, California, and practicing in the fields of erosion and sediment control, watershed restoration, stream restoration, and bioengineering. His career has focused on practical solutions applicable in the arid and semi-arid conditions of the Sierra Nevada Mountains, emphasizing hands-on methods and guidance to contractors. He brings over 21 years of experience training others in erosion and sediment control, watershed restoration, stream restoration, and bioengineering through multiple channels including online video trainings. He has been an adjunct instructor at Shasta College since 1998, teaching Watershed Restoration and Construction Site Erosion Control.

Drawing on his training and experience as a Watershed Geologist, he has worked to support effective stabilization and revegetation on complex construction projects in CA and elsewhere for more than three decades. His projects have tackled saprolitic bedrock materials, creeks populated with endangered species, and sites where prior construction measures have failed. McCullah authored the first Erosion and Sediment Control BMP Manuals for Shasta County in 1992 and then published the award-winning Erosion Draw Manual with AutoCad Typical Drawings in 1994. As a member (and instructor) with the International Erosion Control Association (IECA), John also served on its Board of Directors for 5 years. He served as co-author on the report for the National Cooperative Highway Research Program on "Environmentally Sensitive Channel and Streambank Protection Measures."

John managed the Caltrans 24-hour Certified Stormwater Training at the College and is the Project manager for the Shasta College Erosion Control Training Facility, a six-acre site built to replicate construction site conditions in order to research and demonstrate the proper use of BMPs. For over six years John was the co-instructor for "Field Application Training for Erosion and Sediment Control BMPs on Caltrans Construction Sites", two-day field courses provided to personnel at each

Caltrans District. This hands-on training led John to develop and produce the erosion control training video series "Dirt Time" with John McCullah. It has been said that "Dirt Time is to Erosion Control as *This Old House* is to home renovation". Salix has also developed the popular design manuals on CD; ErosionDraw, BioDraw, and ESenSS.

Mr. McCullah has been a Certified Professional in Erosion and Sediment Control (CPESC #311) since 1986, has a BS in watershed geology from Humboldt State University, and is a California Contractor.

INTRODUCTION

Bioengineering is steadily gaining in popularity. Its practice is seen as both traditional and at times quaint and old fashioned, as well as cutting-edge and technically advanced. The practice of bioengineering or biostabilization has been around a long time. While remaining rooted in tradition and empirical performance, the field has evolved to incorporate modern construction materials, improved science related to design specifications, and broader targeted features and benefits. The treatments found within this book range from familiar and ancient measures, to creative, sophisticated, and even whimsical measures (such as "live smiles"), each suitably adapted to the context of its site and project goals.

The authors conceived of a book featuring bioengineering project case studies to help various professionals involved in planning and design to understand how bioengineering has been successfully used to meet multifaceted project goals. The range of examples and level of detail provided were chosen to demonstrate the decision making that was considered in the adoption and tailoring of ecologically rich and technically sound measures. The projects presented feature treatments that allow sites to function with high resilience and sustainability and with low maintenance inputs. We hope this book will succeed in helping more people recognize when to consider using biotechnical and bioengineering measures in a wide range of settings.

The case studies included in this volume reflect many variations of bioengineering treatments that are able to stabilize soils, promote habitat enhancement, and protect water quality. The most common venue for selecting a biotechnical or bioengineering approach is along highways, adjacent to lakeshores, rivers, and for upland slope stabilization. Restoration, rehabilitation, and mitigation projects in these venues often rely on bioengineering elements. Increasingly, traditional civil engineered projects are incorporating biotechnical design criteria. Biotechnical techniques are frequently employed when two or more, equally important, project goals are identified, e.g., when a highway embankment adjacent to a river or stream must be stabilized AND the sensitive aquatic habitat must be preserved and/or enhanced. Another example might be a land development project that must satisfactorily address issues of sustainability, hydro-modification, esthetics, and local biodiversity. It is noteworthy that the reduction or elimination of adverse project hydromodification is a requirement of the Clean Water Act.

NCHRP Report 544-Environmentally Sensitive Channel and Bank Protection Measures (McCullah and Gray 2005) details over 50 biotechnical/bioengineering methods. This technical report resulted from a problem statement developed by State DOT Highway Engineers that read as follows: "Are there any alternatives or additions (soft engineering practices) that can help the highway engineer design projects that both protect critical infrastructure while restoring ecological health." A large number of stream bank stabilization techniques were considered for inclusion in the report; however, many were excluded because they offered little in the way of environmental benefit beyond simply reducing bank erosion and channel incision. To be included in these guidelines, the techniques had to protect or enhance aquatic or terrestrial habitats, provide esthetic value, or both. Today bioengineering practices often must be considered and effectively implemented in order to obtain environmental permits. Accordingly, this case studies book may provide useful ideas and possible solutions for design professionals, resource agency personnel, and community advocates alike.

W. Goldsmith et al., *Bioengineering Case Studies: Sustainable Stream Bank and Slope Stabilization*,
DOI 10.1007/978-1-4614-7996-3_1,

Defining Bioengineering

A host of terms has been used interchangeably throughout this book to describe similar treatments as captured in the case studies within this publication. *Ingenieurbiologie* was the term used in German, perhaps best translated as "engineered biology," that appeared first in a mid-nineteenth century publication (Goldsmith and Bestmann 1992). The term "water bioengineering" was adopted by Schiechtl and Stern (1994, 1997) to describe water course and shoreline protection techniques.

Charles Kraebel (1935), a landscape architect with the USDA Forest Service, published a circular entitled Erosion Control on Mountain Roads in which he adopted some precursor bioengineering terms, e.g., a "wattle." This term eventually became synonymous with a "fascine."

Somewhat confusingly, in the USA bioengineering is a term used often to label a field of expertise in biologic and genetic engineering that is related primarily to medicine and healthcare. A quick check of most dictionaries will establish this prevailing definition. Few definitions refer to "bioengineering" as the application to environmental modifications such as erosion control, slope stabilization, and stream bank protection. Commonly used terms for this purpose include:

Soil Bioengineering

Bio-stabilization

Biotechnical Erosion Control

Biotechnical Slope Protection

Ecological Engineering

Green Construction

The authors have decided to settle on "bioengineering" in the title of the book as the closest, and also the simplest, English translation of the German language earliest use of the term "*ingenieurbiologie*." To further explain and refine the scope of the term, additional explanation can be added, as in this book featuring stream bank and shoreline protection, watershed restoration, and slope stabilization case studies. Use of woody species is traditional, but increasingly the value of grasses and other herbaceous plants is understood (Goldsmith 2006).

Although conventional engineering techniques (notably riprap and other inert materials) represent solutions to bank stability issues that have easily quantifiable responses to applied hydraulic and geotechnical forces, bioengineering techniques can often supplant or augment the current conventional engineering materials. Bioengineering in the riverine environment (because of the tractive forces below the annual high water) often dictates the use of bioengineering elements combined with structural/mechanical elements. The use of vegetation in combination with structural elements can stabilize and reinforce stream banks and critical slopes by functioning in a mutually reinforcing manner. Structural components used can include stone, rock, concrete, timber, wire, and/or geosynthetics. The use of these elements in combination with vegetation is now well established. This approach is often referred to as "biotechnical stabilization," although many practitioners use the terms interchangeably and do not differentiate.

Soil bioengineering involves the use of vegetation, primarily live cuttings, which are imbedded in the ground in various arrays where they act as soil reinforcements, hydraulic drains, wicks, and barriers to surficial erosion or shallow earth movement (Gray and Sotir 1996). Generally, live cuttings are harvested and installed in the ground in a dormant condition, where they sprout new roots and shoots when temperature and moisture on site triggers growth and establishment. Plants can add to soil consolidation by contributing suction forces to the soil matrix, and conversely excessively compact soil hinders vegetation, making soil compaction a key factor for project success (Goldsmith et al. 2001).

For ease of reference, this book includes an appendix with definitions and thumbnail illustrations of treatment measures common in bioengineering and biotechnical stabilization practice. The case studies also include several additional variants and/or newer measures. These are identified in a separate glossary in the appendix.

Sustainability Considerations

Sustainability may not have been an explicit goal in all the case studies described herein, but it was an implicit goal. If sustainability is a stated project goal, then bioengineering will be a useful approach to adopt and implement. Perhaps the most rudimentary example to illustrate the need for an alternative approach is the widespread use of riprap or concrete armor for stream bank stabilization. Rock armor or riprap is often the first choice in many cases where stream bank erosion and scour jeopardize built structures, but where a bioengineering alternative can offer distinct advantages.

Rivers are dynamic systems whose morphology depends on the interaction between fluid flow and the materials present at the channel boundary. Channels adjust to changing environmental conditions or disturbances, be they natural or anthropogenic, by varying characteristics of their geometry, hydraulics, and sediment load. In terms of channel geometry, bed degradation and/or bank widening represent important mechanisms of channel response and energy dissipation in alluvial channels. These processes of channel adjustment can lead to bank over-steepening, geotechnical instability, and bank retreat, often requiring stabilization of the banks in order to protect land and infrastructure. Evaluating the physical characteristics of the river flows and bank stability parameters is complicated in itself, and standard inert material treatments require rigorous analysis and design, even for effective riprap and similar treatments. However, these materials often fall short, by essentially covering banks with barren or hard materials that can be observed to remain barren or become populated by invasive species for decades. These conditions are not stable, nor sustainable, in the long run.

Without stability of landscape-scale systems in the form and function, sustainability cannot be achieved. Since the beginning of civilization, land uses from agriculture to urbanization significantly modified and impacted the landscape and the ecological functions we all ultimately depend on. Regardless of whether the responses are immediate or clearly evident to the untrained, they do exist. At times effects may be tolerable or difficult to discern over an extended time period (e.g., the decline in air quality since the dawn of industrialization). At other times, effects can be catastrophic (e.g., erosion and flooding caused by increased peak flow of storm water runoff from paved/roofed areas). Decline and degradation result, and eventually a need arises to manage the system to compensate, mitigate, or repair.

Sustainable solutions to perceived problems are best achieved when based on a comprehensive evaluation of the causes, effects, and relationships of the system as a whole, rather than focusing on the symptom. For example, treatments that allow rainfall to be captured on site, rather than directed on an accelerated basis towards rivers, are more effective than efforts to enlarge streams for flood conveyance. Ideally design teams include scientists and natural resource managers trained and experienced at identifying effective means of achieving sustainability, rather than allowing the design process to be driven by custom and habit.

Cost–Benefit Considerations

When contemplating a biotechnical treatment, there are often cost and benefit questions to consider. These questions span the planning and design process, as well as the construction phase and post-construction maintenance needs. Larger issues may include benefits to esthetics and to natural system functions that garner support from stakeholders, thus allowing review processes to proceed quickly and with positive results. An initial consideration that may arise are additional costs related to the characterization of site conditions in a way that allows understanding of biological factors, as well as the engagement of design team members with expertise related to bioengineering measures.

Costs

Direct projects costs are associated with providing labor, materials, and services. Some of the case studies in this book include a section on costs and costs that are specific to that particular project. Construction costs are relatively direct and straightforward to calculate, e.g., a willow branch might cost $2.00 (Washington State 2002) to be harvested, handled, prepared (soaked or cold storage), delivered to site, and then installed. Many environmentally sensitive stream bank stabilization methods (McCullah and Gray 2005) also include vegetated rock toes or have redirective features such as rock vanes. These combined and/or integrated approaches require a different way of estimating expected costs.

There are several sources documenting the costs of common bioengineering components (plant materials) and techniques. From these one can sometimes extrapolate the cost of the "biotechnical technique" (plant material and structural/engineering materials) by simple addition. For instance, BioDraw 3.0 (McCullah 2004) documents the cost of Brushlayering with Soil Wraps (aka Vegetated Mechanically Stabilized Earth) constructed with woven coir fabric at \$2–\$3/yd^2 plus \$10–\$15/linear ft for the willow branches (Brushlayering). As documented in NCHRP Report 544, the cost of live staking "ranges from $1.50 to $3.00 per stake, including harvesting, transportation, storage, and installation. Typically, costs are closer to $1.50 per stake but may be higher if labor costs are especially high or the harvesting location is a long way from the project site. Estimated labor allocations are 2–5 m^2 (20–50 ft^2) of live staking per work hour or approximately 10–25 stakes per hour. These estimates include all preparations."

There is much difficulty in discussing the absolute and direct costs associated with biotechnical projects. In reviewing completed case studies, sometimes the project owners are reluctant to provide the final costs. Likewise with public works-type projects, it is often difficult to separate out the biotechnical elements from the more traditional engineering elements. A rather straightforward example might be the analysis of biotechnical costs associated with vegetated riprap. The cost of riprap is well documented in most regions. The project proponent might then ask "what does it cost to vegetate the riprap with willow poles?" The answer could be quite simple if the willow branch is installed during the construction and placement of the rock.

In essence, the cost of Vegetated Riprap is simply the cost of riprap plus the cost of cutting, handling, delivery, and placement of the willow poles. Having a worker on hand to install the pole as the rock is being placed is minimal as an additional cost to construction. If, however, the design calls for vegetating the riprap after construction, then the costs associated with installation of that willow pole is substantially higher due to the difficulty of augering through the rock and establishing proper contact with the native soil below. The method of design and placement of plant material during construction is highly preferable to installing the plant material as an afterthought, as demonstrated through research (McCullah and Gray 2005). When adding plantings after rock placement, costs increase and performance tends to diminish.

Value

Value added is by nature indirect and more complicated to determine. What is the value, for example, of shade and cover provided to a project adjacent a water body or what is the value in of reduced raindrop erosion, or finally the value of faster and easier permitting? In recent decades, bioengineering has gained preference in the eyes of regulators, due to the water quality and habitat values it provides. For sites with known endangered species, structural measures are frequently forbidden, and site-specific vegetative measures are the sole option acceptable to regulators. Consider for a moment the value of implementing biotechnical stream bank stabilization measures to a Department of Transportation Project manager when State natural resource agencies will not issue project permits without integrating habitat-enhancing techniques into the project design.

In other cases, there may be no clear cut prohibition, but the review of impacts and benefits favors bioengineering approaches. Increasingly, the practice is sought after for reasons that are grounded in the pursuit of sustainable development. Using vegetation to reduce erosion offers the double benefits of controlling the movement of sediment and nutrients into water bodies, and also the increase of soil organic matter, which can greatly compensate greenhouse gas emissions.

The primary value of bioengineering treatments is the use of vegetation, which in turn relies on sunlight as an energy input that drives ongoing growth, regeneration, and self-maintenance. At an essential level, bioengineering establishes conditions that allow stable landforms and healthy soils populated by robust plant communities to harvest solar radiation. These conditions in turn provide a host of functions that benefit water quality and habitat value and also remove and store carbon from the atmosphere. Choosing bioengineering over its common alternatives fundamentally contributes to ecosystem productivity, balance, and resilience for each patch of land where it is applied, accruing benefits at a watershed, regional, and ultimately a planetary scale.

Cited References

Goldsmith, W. (2006). Soil Strength Reinforcement by Plants. IECA conference proceedings. Recipient of Most Distinguished Technical Paper Award.

Goldsmith, W.and Bestmann, L. (1992). An Overview of Bioengineering for Shore Protection. IECA Conference Proceedings.

Goldsmith, W., Marvin Silva, PhD, PE, Craig Fischenich, PhD, PE (2001). Determining Optimal Degree of Soil Compaction for Balancing Mechanical Stability and Plant Growth Capacity. ERDC Technical Note EMRRP.

Gray, D.H. & Sotir, R. (1996). Biotechnical and Soil Bioengineering Slope Stabilization. John Wiley and Sons, New York, N. Y.

Kraebel, C. J., (1935). Erosion Control on Mountain Roads, Department of US Agriculture, Circular 380

Schiechtl, H.M. and Stern, R. (1994). Water Bioengineering Techniques for Watercourse Bank and Shoreline Protection, Osterreichischer Agrarverlag, Klosterneuburg, Austria

Schiechtl, H.M. and Stern, R. (1997). Water Bioengineering Techniques for Watercourse Bank and Shoreline Protection. English Translation with additions. Blackwell Science, Ltd.

McCullah, J. (2004). Bio Draw3.0 Compendium of Biotechnical Soil Stabilization Solutions (CD), Salix Applied Earthcare, Redding, CA

McCullah, J., and Gray, D.H, (2005). NCHRP Report 544 – Environmentally Sensitive Channel and Bank Protection Methods,

Washington State Dept of Fish & Wildlife (2002). Integrated Streambank Protection Guidelines, published in co-operation with Washington Dept. of Transportation and Washington Dept. of Ecology, June 2002.

Great Lakes Area

PROJECT #1: FLEMING CREEK

PROJECT TYPE: **Stream bank repair and protection**

PROJECT SCALE: Small

CLIENT/OWNER: Matthei Botanical Gardens
The University of Michigan

TECHNIQUES EMPLOYED: **Rock vanes, Live pole planting**

GEOGRAPHIC LOCATION: Fleming Creek
Northfield TWP
Ann Arbor, MI

Fig. 1 Fleming Creek rock vanes

GEOMORPHIC SETTING: Fleming Creek is a scenic, meandering stream that flows into the Huron River. It drains an area of 31 square miles, through the townships of Ann Arbor, Superior, Northfield, and the City of Ann Arbor. Fleming Creek runs through the Parker Mill County Park and once provided motive power for that mill. It also flows through the University of Michigan, Matthaei Botanical Gardens. The stream bank remediation site is located adjacent a walking trail along Fleming Creek (Fig. 1).

SITE CONDITIONS AND PROBLEMS: Scour erosion at the outside bend of the stream had removed up to 10 ft of the bank and produced a near vertical, unstable slope. The stream channel was also beginning to encroach into the walking path (see Fig. 2). Erosion and bank scour were caused by high velocity, impinging flows that occurred during storm events.

W. Goldsmith et al., *Bioengineering Case Studies: Sustainable Stream Bank and Slope Stabilization*,
DOI 10.1007/978-1-4614-7996-3_2, © Springer Science+Business Media New York 2014

Fig. 2 Stream bank before treatment

TREATMENT OBJECTIVES AND CONSIDERATIONS: The main consideration was the need to arrest erosion and stabilize the bank in an environmentally sensitive manner. The repair technique selected had to redirect stream flow away from the bank, stabilize the bank, promote a diverse hydrologic regime, and preserve existing aquatic habitat. An approach was needed that would avoid sloping back or regrading the bank. This action would have resulted in additional loss of bank at the top. These considerations and requirements dictated the use of structures placed in a transverse direction to the bank such as groins, vanes, or bendway weirs. Rock vanes were selected as the most suitable alternative to achieve these objectives. The vanes were placed in conjunction with live willow poles that were inserted into the soil along the toe of the bank and in the vane keyways.

TREATMENTS SELECTED

Rock vanes. Elongated, rock prisms with a triangular cross section were placed at strategic locations to deflect or redirect impinging flow away from the eroding bank. The vanes were oriented in an upstream direction at a 30° angle with the bank. The vanes sloped down from the bank to a point at their upstream end. The vanes were submerged along most of their length (except at very low water) and only encroached into the stream channel about one-fourth of the channel width. The vanes were keyed into the bank (by placing rock in a trench oriented perpendicularly to the bank) to prevent flow from cutting around the vanes during high water (see Fig. 3).

Fig. 3 Construction of keyway for a rock vane

Fig. 4 Placement of willow poles in keyway

Live willow pole planting. Live willow poles or cuttings were inserted into the sides of the key way trenches and also along the toe of the bank (see Fig. 4). The cuttings eventually rooted and leafed out. This vegetation provided additional protection against erosion and bank scour.

Rock sill removal. An old rock sill placed on the channel bottom was removed prior to construction of the rock vanes. This sill was oriented in such a way that it actually directed flow into the eroding bank.

OBSTACLES TO IMPLEMENTATION: The rock vanes where installed using volunteer and prison inmates from the county sheriff's department. Initially one of the main obstacles to implementation was lack of familiarity on the part of both regulators and installers with redirect techniques such as rock vanes. Regulators even expressed some skepticism about the technique because of this lack of familiarity and unawareness of the superior benefits provided by flow redirection as opposed to bank armoring. The placement of silt curtains around the site was required in advance of and during construction. However, no evidence of significant turbidity was manifest. The keyway trenches were excavated and rock placed by machine; on the other hand, the rock vanes were placed by hand. The latter was not a serious limitation because of the relatively small size and number of vanes used on this project.

PERFORMANCE EVALUATION: The rock vanes have successfully "bumped" the flow away from the bank towards the center of the channel. This change can be observed by noting the stream flow patterns in the vicinity of the vanes. Scouring and erosion downstream of the vanes have ceased; instead water velocity in these areas is now minimal, and sedimentation has been enhanced. This change has improved bank stability. The vanes have also enhanced aquatic habitat by creating more hydrologic diversity with the formation of scour pools (at the vane tips) and zones of slow and fast water. The willow poles successfully leafed out and rooted thus providing additional bank protection and visual benefits (see Fig. 5).

Fig. 5 Stream bank after treatment

BENEFITS AND LESSONS LEARNED:

1. This project demonstrates the use of an effective, environmentally sensitive alternative to bank armoring. The project is a good example of a redirect approach to stream bank repair that preserves the bank line without the need for regrading.

2. The project provided hands-on opportunities for volunteer participation. The work was accomplished by well supervised, volunteer labor including some help from prison inmates in the county sheriff's department.

ADDITIONAL READING:

NCHRP (2005). Environmentally Sensitive Channel and Streambank Protection Measures, NCHRP Report 544. Guidance manual prepared for the Natl. Coop. Hwy. Research Program, Washington, DC.

PROJECT #2: GATEWAY GARDEN

PROJECT TYPE: **Runoff and erosion control**

PROJECT SCALE: Small

CLIENT/OWNER: Nichols Arboretum
The University of Michigan

TECHNIQUES EMPLOYED: **Stepped pools, riparian planting**

GEOGRAPHIC LOCATION: Western entrance
Nichols Arboretum
Ann Arbor, MI

Fig. 1 View of Gateway Garden

GEOMORPHIC SETTING: The Gateway Garden sits at the head of a large ravine in a glacial moraine (Fig. 1). The ravine is part of the Huron River drainage that flows into Lake Erie. The garden is located in a 17-acre sub watershed that includes a cemetery and roads that generate significant amounts of runoff. An intermittent stream that flows through the garden drains a watershed of approximately 33 total acres that includes not only the cemetery but also roads, parking lots, and building roofs. The stream is fed entirely by storm water runoff, so it fluctuates between being nearly dry and discharging large amounts of water.

SITE CONDITIONS AND PROBLEMS: High runoff from impervious surfaces in the surrounding watershed combined with direct discharges from street drains have resulted in large storm flows and serious erosion problems. Uncontrolled storm flows caused both channel head-cutting and down-cutting at the head of the ravine. These uncontrolled flows also impeded access to the Interpretive Center located next to the west entrance to the Arboretum (see Fig. 2). Soils at the site are primarily sandy to clay loams over a hard clay till, which is exposed at the bottom of the ravine. Areas with sandy to loamy soils are highly erodible when vegetation is removed.

W. Goldsmith et al., *Bioengineering Case Studies: Sustainable Stream Bank and Slope Stabilization*,
DOI 10.1007/978-1-4614-7996-3_3, © Springer Science+Business Media New York 2014

Fig. 2 Uncontrolled runoff from cemetery above treatment site

TREATMENT OBJECTIVES AND CONSIDERATIONS: An ancillary objective would have been to work on source reduction, i.e., to reduce or divert the amount of runoff entering the head of the ravine. For several reasons this option was not available. Instead a variety of different techniques were implemented, to correct and mitigate existing erosion and runoff problems. Although source reduction was not a project option, some effort was directed towards detaining and controlling runoff entering the head of the ravine. This was accomplished mainly by constructing a series of landscaped, "stepped pools" (low-head boulder drops). The pools and connecting channels detain runoff, dissipate energy, and allow for some infiltration of runoff into the ground. Riparian plantings provide additional erosion protection, phyto-remediation, and attractive landscaping.

TREATMENTS SELECTED:

Stepped pools. A series of low-head, boulder drop structures connected by a low-gradient channel were constructed to handle and slow runoff flowing into the head of the ravine (see Fig. 3). The services of a professional engineering firm and earthwork contractor were engaged to design and construct the drop structures. The calculated peak 10 and 100-year recurrence interval flow rates for these pools are 5.0 and 16.5 cfs, respectively. An outlet drain conveys additional storm runoff from the adjacent Washington Heights sub-watershed with a drainage area of approximately 1.3 acres. This drainage area includes mostly impervious roadway, roof surface, and parking lots. The channel between pools was designed to carry the 100-year peak flow at a non-erosive velocity. A triangular channel at a 1 % slope was selected to minimize earthwork and provide a subcritical channel design. Based on Manning's equation, the maximum velocity in the channel during a 10-year event will be 2.6 fps. A minimum of Type B vegetation (NRCS) was required to adequately protect the channel against scour erosion.

Fig. 3 Construction of stepped pools

Fig. 4 Riparian plantings around stepped pools

Riparian planting. The edges of the pools and connecting channels were landscaped with a variety of riparian plants that provided phyto-remediation and additional erosion protection. A variety of prairie and river's edge species were planted such as river bank sedge (*Carex emoryi*), prairie cordgrass (*Spartina pectinata*), Gray's sedge (*Carex grayii*), Indian hemp (*Apocynum cannabinum*), blue flag iris (*Iris virginica*), and cupplant (*Silphium perfoliatum*). Calculations showed that the allowable velocity in a channel with this type of vegetative lining is 6.6 ft/s. Accordingly, a Type B vegetative cover will provide adequate stabilization in the channel. The riparian plantings were carried out by supervised volunteers (see Fig. 4).

Fig. 5 Upper stepped pools—1 year after construction

OBSTACLES TO IMPLEMENTATION: Skepticism on the part of some about the effectiveness of the proposed treatment to control runoff and provide unimpeded access to the Interpretive Center was perhaps the main obstacle to implementation. A considerable amount of persuasion was required to avoid implementing a remedy that simply collected the runoff coming from the cemetery and road system and piping it directly to the ravine.

PERFORMANCE EVALUATION: The repair work has been tested by several large storm events since construction. Head cutting has been entirely stopped by the landscaped, stepped-pool system. This area that once experienced ankle-deep runoff during storms (refer to Fig. 2) is now covered with riparian vegetation which further slows runoff. The stepped-pool and riparian plantings provide an attractive landscaping benefit (see Fig. 5). The project received an award from the American Society of Landscape Architects.

BENEFITS AND LESSONS LEARNED:

1. Runoff control and erosion prevention at the head of the ravine were achieved by a combination of environmentally sensitive techniques that included low-head boulder drops (stepped pools) and vegetative riparian plantings.

2. Although source reduction was not a project option, it was possible to detain and control runoff entering the head of the Glen by constructing a series of landscaped, stepped pools.

3. A liability was converted to an asset by controlling runoff and using the water to help support and maintain riparian plantings and landscaping.

ADDITIONAL READING:

Gray, D.H., Grese, R., and Orlow, T. (2004). Saving School Girl's Glen. *Erosion Control*, Vol. 11, No. 2, pp. 34-45

Ayres, Lewis, North & May (2002). Reader Center Stormwater Channel Design. Report prepared for the University of Michigan. Proj. No. 183752-01, February 2002

PROJECT #3: SCHOOL GIRL'S GLEN

PROJECT TYPE: **Watershed restoration**

PROJECT SCALE: Medium

CLIENT/OWNER: Nichols Arboretum
The University of Michigan

TECHNIQUES EMPLOYED: **Vegetated MSE,**
Stone weirs
Live staking and native plantings

GEOGRAPHIC LOCATION: Western entrance
Nichols Arboretum
Ann Arbor, MI

Fig. 1 Gabion check dam

GEOMORPHIC SETTING: The ravine, known locally as School Girls Glen, is part of the Huron River drainage that flows into Lake Erie. High runoff from impervious surfaces in the surrounding watershed combined with direct discharges from street drains resulted in large storm flows and serious erosion problems.

SITE CONDITIONS AND PROBLEMS: Uncontrolled storm flows caused channel head-cutting and down-cutting in the Glen itself. These in turn under-mined adjacent stream banks and side slopes, which resulted in serious slope failures. The combination of surficial erosion, channel scour, and slope failures produced massive losses of soil (see Fig. 2). At low water, a large sediment delta was clearly visible at the confluence of the Glen's ravine with the Huron River. Soils are primarily sandy to clay loams over a hard clay till, which is exposed at the bottom of the ravine in the Glen. The areas with sandy to loamy soils are highly erodible when vegetation is removed. One of the factors contributing to erosion of the steep slopes is the prevalence of nonnative invasive species that have shaded out the native under story vegetation and prevented regeneration of the native tree species.

W. Goldsmith et al., *Bioengineering Case Studies: Sustainable Stream Bank and Slope Stabilization*,
DOI 10.1007/978-1-4614-7996-3_4, © Springer Science+Business Media New York 2014

Fig. 2 Slope failure above check dam site

TREATMENT OBJECTIVES AND CONSIDERATIONS: The main goal was to arrest erosion and slope failures that were occurring in the ravine and to eliminate sediment discharges to the Huron River. An ancillary objective was to carry out this remedial work as a demonstration project using environmentally sensitive stream bank protection and erosion control methods. A variety of different techniques were considered to correct and mitigate problems in the large ravine that comprised the lower part of the Glen. Measures were implemented to better handle street drain discharges which had initiated tributary gullies on the side slopes of the Glen, to repair slope failures on the ravines sides, and to prevent channel scour and erosion on the ravine bottom. The techniques that were selected were relatively low-cost, environmentally sensitive measures that could be implemented using volunteer labor.

TREATMENTS SELECTED:

- **Live staking and native planting.** Live staking and transplants (see Fig. 3) were used to reestablish vegetation and help stabilize selected side slope areas in the ravine. The live stakes, mostly willow (*Salix* spp.), were cut in Nichols Arboretum or nearby natural areas along the Huron River.

- **Stone weirs.** Naturally occurring boulders were used to construct a series of weirs or low-head drop grade stabilization structures (see Fig. 4) at critical locations along the channel. Artificial boulder cascades that mimicked natural cascades were also constructed immediately downstream the channel drops to dissipate energy of flowing water.

Fig. 3 Transplants and live cuttings

Fig. 4 Artificial, constructed boulder cascade

- **Vegetated mechanically stabilized earth repair fill.** A large cavity in the ravine side slope was filled and the bank restored to its original contour by backfilling with layers of sand wrapped (reinforced) with recycled, fiber-backed carpeting in lieu of a geotextile fabric (see Fig. 5). Live stakes were inserted in and around the reinforced repair fill.

- **Gabion check dam.** A low check dam was constructed in the channel at a critical location to arrest channel down-cutting and stabilize the toe of adjacent unstable side slopes in the ravine (refer to Figs. 1 and 6).

Fig. 5 Construction of VMSE plug to fill void in slope

Fig. 6 Construction of gabion check dam in ravine

- **Invasive plant removal.** Invasive species were targeted for removal during the project to promote the establishment of native groundcovers and to enhance vistas into the Glen.

Fig. 7 VMSE slope repair after construction

PERFORMANCE EVALUATION: The repair work has been tested by several large storm events since construction. Channel down cutting has been greatly minimized, and the ravine side slopes are now relatively stable (see Fig. 7). Removal of invasive plant species has allowed native understory and ground cover vegetation to reestablish in the Glen.

OBSTACLES TO IMPLEMENTATION: A considerable amount of persuasion was required to avoid implementing a remedy that simply collected the runoff coming from the cemetery and road system above the Glen and piping it directly to the ravine without concern for erosion problems that would continue to occur there. Access to a deep and well-incised ravine also posed a challenge. This obstacle limited use of heavy equipment and required extensive use of manual labor for implementing corrective measures.

BENEFITS AND LESSONS LEARNED:

1. The repair and restoration of the Glen were achieved by a combination of techniques ranging from mainly vegetative, e.g., removal of invasive species, to primarily structural, e.g., check dam and channel drop structures.

2. The work was accomplished mainly by well supervised, volunteer labor. The final product has served as a highly visible erosion control demonstration project that has provided innumerable hands-on opportunities for students and community volunteers.

3. The erosion control and restoration work in the Glen are showcased via a "Clean Water Trail" with numbered markers at selected viewing locations.

ADDITIONAL READING:

Gray, D.H., Grese, R., and Orlow, T. (2004). Saving School Girl's Glen. *Erosion Control*, Vol. 11, No. 2, pp. 34- 45

Gilzow, S. (2009). Soft Healing at the Arb. The Ann Arbor Observer (2/26/09) http://arborweb.com/articles/soft_healing_at_the_arb_14.html

PROJECT #4: RIVER LANDING

PROJECT TYPE: **Stream bank repair and stabilization**

PROJECT SCALE: Medium

CLIENT/OWNER: The University of Michigan

TECHNIQUES EMPLOYED: **Vegetated riprap, Live fascine, Erosion control blanket (ECB), Re-vegetation**

GEOGRAPHIC LOCATION: Nichols Arboretum
Huron River
Ann Arbor, MI

Fig. 1 Stream bank protection measures

GEOMORPHIC SETTING: The Huron River which drains into Lake Erie has a drainage area of 730 sq. miles, a daily mean stream flow of 380 cfs, and a maximum recorded peak discharge at the site of 3,330 cfs (1968). Dams immediately upstream and downstream of the site tend to regulate flow and suppress peaks in this reach of stream. The river is also quite wide at the site which tends to decrease velocity and depth of flow.

SITE CONDITIONS AND PROBLEMS: The site in question was one of the most heavily used and degraded sites in the University of Michigan Arboretum. It traditionally provides one of the main access and viewing points on the river. Bank erosion was a perennial problem at this site. Concrete rubble was placed on the bank in an attempt to control erosion. The rubble was unsightly and largely ineffective (see Fig. 2). Effective repair and restoration required removal of the rubble and replacement with a more natural bank protection system.

W. Goldsmith et al., *Bioengineering Case Studies: Sustainable Stream Bank and Slope Stabilization*,
DOI 10.1007/978-1-4614-7996-3_5, © Springer Science+Business Media New York 2014

Fig. 2 Stream bank before restoration

TREATMENT OBJECTIVES AND CONSIDERATIONS: The main goal was the need to remove the concrete rubble on the bank and replace with a natural, less visually intrusive protection system. Another goal was to provide visitor access to the river at the site. These considerations resulted in a bank protection system consisting of a vegetated riprap at the toe of the bank combined with a fascine row above the rock armor and an erosion control blanket placed above the fascine to the bank top. The concrete rubble was removed and the bank graded back to a stable 2:1 (H:1) slope angle. Access to the river had to be provided as part of any restoration bank protection system.

TREATMENTS SELECTED:

A combination of treatments and repair measures that were compatible with site conditions were selected to meet the stabilization requirements and goals. Individual measures consisted of the following:

Vegetated riprap. The bank toe was armored with riprap. The riprap was vegetated using the joint-planting method, viz., live stakes were inserted between openings or interstices in the rock armor (see Fig. 3).

Live fascine. A single live fascine was placed above the rock armor. The bottom of the erosion control blanket was run into the fascine trench, and the fascine was anchored in the trench using both live stakes and inert construction stakes (refer to Figs. 1 and 4).

Fig. 3 Inserting live stakes in riprap

Fig. 4 Live fascine staked over ECB at toe of stream bank

Erosion control blanket. An erosion control blanket (ECB) was placed on the bank and up over the top (see Fig. 5). The blanket was anchored to the bank using staples and pegs. Topsoil was placed and spread over the bank before placing the erosion control blanket.

Fig. 5 ECB rolled in place over topsoil on stream bank

Revegetation. A seed mix of native grasses and plants was spread on the bank before placing the erosion control blanket.

OBSTACLES TO IMPLEMENTATION: Some concern was expressed about the bank vegetation limiting access to the river. This concerned was allayed by constructing an attractive set of stairs consisting of granite blocks from the top of bank to the water's edge. In addition the top of bank was planted to grass and picnic tables provided so visitors could enjoy riverine views.

PERFORMANCE EVALUATION: Vegetative establishment beneath the erosion control blanket and in the fascine have generally been excellent. The bank is now stable (see Fig. 6). The live stakes in the riprap have not fared as well. A year after installation only about 30 % of the live stakes have survived and sent out shoots. The rock armor has survived several high water events but has not been tested by floods approaching peak flows of record. Attractive steps built into the bank at two locations in the protected reach now provide ready access to the waters edge.

Fig. 6 Stream bank 2 years after treatment

BENEFITS AND LESSONS LEARNED:

1. This project is a good example of an environmentally sensitive, stream bank repair approach that also provides access to the river. The project has provided hands-on opportunities for volunteer participation and training.

2. The project resulted from planning and cooperation on the part of personnel from the University of Michigan, the City of Ann Arbor, Washtenaw County Road Commission, a private consulting firm (ECT Inc.), and the US Natural Resources Conservation Service.

3. Revegetation of the rock armor by the insertion of live stakes through interstices in the riprap has only been partially successful. The bent willow pole or willow bundle method, which eliminates many of the problems associated with the joint planting method, should be considered in the future.

ADDITIONAL READING:

USDA/NRCS (2005). Workshop on Streambank Restoration Using Natural Materials and Soft Engineering. *Workshop* held at the Nichols Arboretum, University of Michigan, 10/28/05

NCHRP (2005). Environmentally Sensitive Channel and Streambank Protection Measures, NCHRP Report 544. Guidance manual prepared for the Natl. Coop. Hwy. Research Program, Washington, DC.

PROJECT #5: NICHOLS DRIVE

PROJECT TYPE: **Streambank and road stabilization**

PROJECT SCALE: Medium

CLIENT/OWNER: The City of Ann Arbor
The University of Michigan

TECHNIQUES EMPLOYED: **Rock vanes, Vegetated MSE, Chimney drain, Re-vegetation**

GEOGRAPHIC LOCATION: Nichols Drive
Huron River
Ann Arbor, MI

Fig. 1 Nichols Drive road embankment

GEOMORPHIC SETTING: Nichols Drive is an unpaved, riverine road that is supported atop a streambank next to a sharp outside bend in the Huron River (Fig. 1). The streambank lies at the base of a steep glacial moraine. The Huron River has a drainage area of 730 sq. miles, a daily mean stream flow of 380 cfs, and a maximum recorded peak discharge at the site of 3,330 cfs (1968). Dams upstream and downstream of the site now tend to regulate flow and suppress peaks in this reach of stream. The river is relatively narrow at the site which tends to increase velocity and depth of flow.

SITE CONDITIONS AND PROBLEMS: The river bank at the site consists of fill soils that were placed to support an unpaved roadway. The roadway is about 10–12 feet above the level of the river. A timber retaining wall was constructed at the toe of the fill to help keep it in place. The remains of this wall were uncovered during exploratory borings and excavation. Seepage from a steep slope uphill of the road and from a drainage ditch entered the soil beneath the road. A ditch was located along the south (uphill) edge of the roadway to collect and divert runoff into small culverts and into the river. These undersized culverts tended to plug with debris which

DOI 10.1007/978-1-4614-7996-3_6,

Fig. 2 Cracking and slumping in streambank

allowed water to pond in the ditch and seep into the soil beneath the roadway thereby saturating and weakening the road prism. This seepage saturated and weakened the soil causing the edge of the road prism to slump towards the river as shown in Fig. 2. Stream scour and erosion at the toe of the riverbank also decreased lateral support and exacerbated the slope failure. Effective repair and stabilization required addressing both these problems.

TREATMENT OBJECTIVES AND CONSIDERATIONS: The main consideration was the dual need to stabilize the riverbank against slope failure and to protect the toe against stream erosion. It was also necessary to raise the roadway surface a couple of feet and displace the road laterally (into the hillside) about 5 feet. Some type of retaining structure or system was required to support the earthen bank fill and road prism. Subsurface drainage measures were also needed to intercept seepage and prevent the fill beneath the road from becoming saturated.

Prevention of continued erosion/scour at the toe of the riverbank required the use of either bank armor and/or redirective measures such as rock vanes. The latter approach is more environmentally friendly because it displaces flow away from the bank towards the center of the stream, provides greater flow diversity, and promotes sediment deposition between the vanes. The site is highly visible, and many visitors use the road to access the Arboretum. One of the treatment goals was to implement permanent repair and stabilization measures that are environmentally sensitive and visually attractive.

TREATMENTS SELECTED:

A combination of treatments and repair measures that were compatible with site conditions were selected to meet the stabilization requirements and goals. Individual measures consisted of the following:

- **Vegetated mechanically stabilized earth (VMSE).** A VMSE retaining structure with a 1:1 (H:V) sloping face was selected to support the roadway. Mechanically stabilized earth (MSE) consists of horizontal lifts (layers) of earth that are wrapped and reinforced with geotextiles and/or geogrids (see Fig. 3). In the case of VMSE live, cut branches are also inserted between successive lifts (NCHRP 2004). A rock footing was placed in an excavated trench beneath the VSME structure.

- **Rock vanes.** Elongated, rock prisms with a triangular cross- section were placed at strategic locations to deflect or re-direct impinging flow away from the river bank (see Fig. 4). The vanes are oriented in an upstream direction at a 30-degree angle with the bank. The vanes were keyed into the bank (by placing rock in a trench oriented perpendicularly to the bank) to prevent flow from cutting around the vanes during high water.

Fig. 3 Installation of VMSE (soil lifts reinforced w/ geogrids and cut branches)

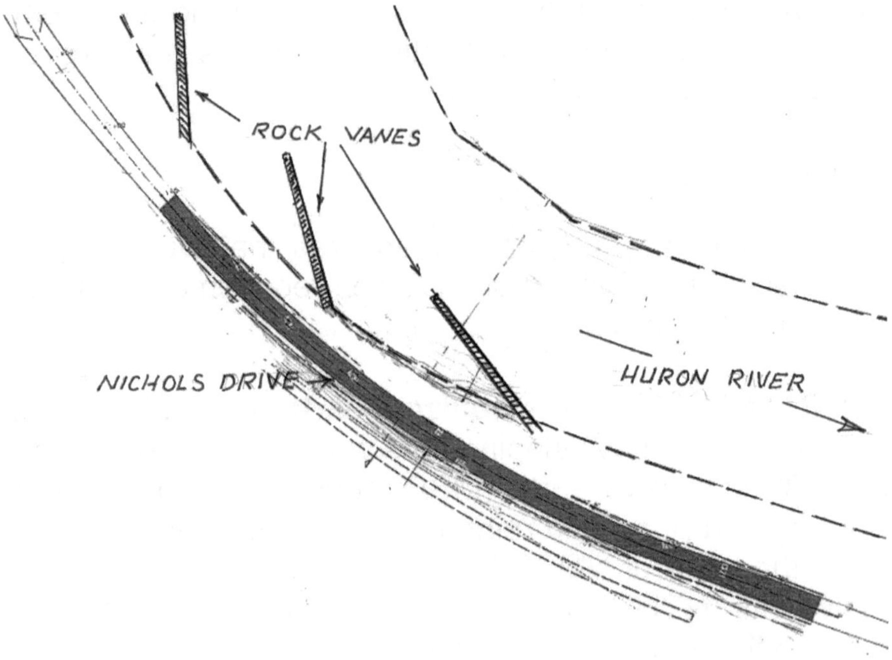

Fig. 4 – Schematic diagram of rock vane installation

- **Surface and subsurface drainage.** A drainage ditch was excavated along the road to intercept upslope runoff. A drainage pipe was also placed in a trench excavated along the road to convey slope runoff away from the site. A sloping drainage course (chimney drain) consisting of a polymeric blanket drain with a high in-plane, hydraulic conveyance was placed behind the VMSE retaining structure. This drainage blanket intercepts seepage and conveys it to a subdrain system that discharges directly to the river. This measure minimizes undesirable saturation of the fill beneath the roadway and structural fill in the VMSE retaining structure.

- **Re-vegetation.** Grass seeding, live staking, and woody transplants were used in addition to the live, cut branches placed between horizontal lifts in the VSME structure. The grass seed mix was placed by hydroseeding the top of the streambank and also directly on the face of the VMSE structure. Small trees were planted along the crest of the streambank next to the roadway. Erosion control blankets were also placed in this area to protect the slope against surface erosion and promote vegetative establishment.

OBSTACLES TO IMPLEMENTATION: Earlier objections and concerns were expressed about a possible adverse influence of the vanes on recreational canoeing. These concerns, however, soon abated. The vanes do not encroach far into the channel, are well submerged along most of their length, provide quiet slack water downstream, and tend to redirect floating objects (e.g., canoes) along with the flow. The rock vanes have effectively redirected high velocity flow away from the project site in spite of the fact they are almost completely submerged and barely visible during high flow events.

PERFORMANCE EVALUATION: The riverbank and road have been stabilized successfully by environmentally sensitive techniques. The site looks visually attractive and can be observed by the public who use the road frequently to access the Arboretum (see Fig. 5). A vegetated mechanically stabilized earth (VMSE) retaining structure was constructed that supports the roadway. A vertical drainage blanket or chimney drain now intercepts seepage that had previously destabilized the stream-bank and road prism. Impinging flow at the toe of the riverbank has been redirected away by the rock vanes, and stream erosion has been completely eliminated.

Fig. 5 Project site 1-month after completion—April 2008

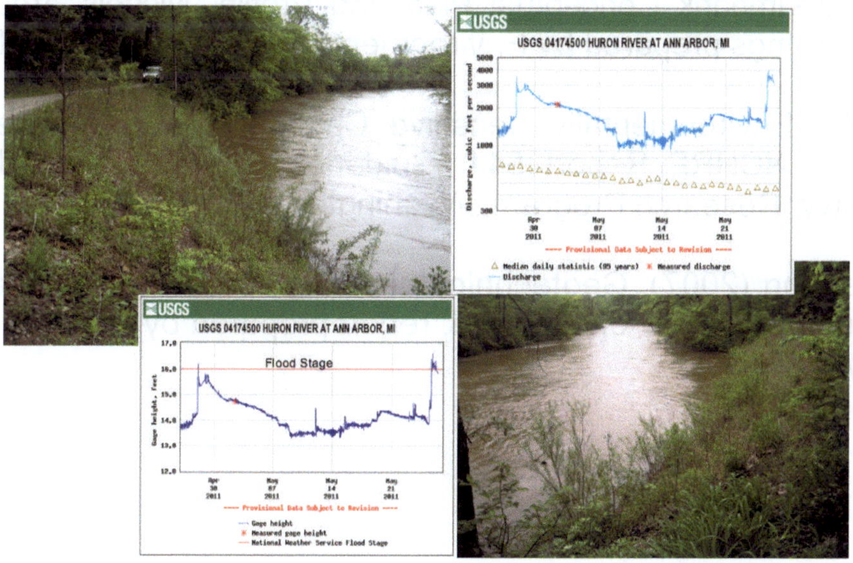

Fig. 6 Project site during bankfull flow—May 2011

The streambank has withstood several bankfull flow events without any damage (see Fig. 6). Vegetation has reestablished well on the river bank as a result of hydroseeding and planting small trees and other vegetation near the top of the river bank next to the road (refer to Figs. 5 and 6).

BENEFITS AND LESSONS LEARNED:

1. This project is a good example of stream bank repair and stabilization that employs several environmentally sensitive measures. It makes an excellent, easily accessible, and highly visible demonstration project.

2. Construction of the VMSE retaining structure required installation of a temporary sheet pile wall in the river channel so that the toe of the slope could be excavated and dewatered. This was done to accommodate a stone footing placed in a trench that was excavated below the water level in the river.

3. Vegetation has reestablished well on the river bank as result of hydroseeding and planting small trees and other vegetation on the crest of the riverbank next to the road and on the slope above the rock vanes.

4. Establishment of live stakes and live cut branches used in the VSME structure has not fared well (<30 % survival). The cuttings were held too long in stagnant water in plastic barrels and inserted into the ground well after the end of dormancy. In addition the stakes were not buried deep enough (at least 2/3 → 3/4 of the stake length should be in the ground).

ADDITIONAL READING:

ECT (2007). Streambank Conceptual Plan and Profile, Nichols Arboretum Access Road, drawings prepared by Environmental Consulting Technologies

NCHRP (2005). Environmentally Sensitive Channel and Streambank Protection Measures, NCHRP Report 544. Guidance manual prepared for the Natl. Coop. Hwy. Research Program, Washington, DC.

Somat Engineering (2007). Geotechnical Investigation for Slope Stability, Nichols Arboretum, University of Michigan, report prepared by Somat Engineering, 17 January 2007

PROJECT #6: HARVARD ROAD

PROJECT TYPE: **Gully repair and watershed restoration**

PROJECT SCALE: Small

CLIENT/OWNER: The City of Ann Arbor
The University of Michigan

TECHNIQUES EMPLOYED: **Rocked pools, Sinuous channel, ECBs, Live staking**

GEOGRAPHIC LOCATION: Nichols Arboretum
University of Michigan
Ann Arbor, MI

Fig. 1 Stepped, rocked pools

GEOMORPHIC SETTING: Broad sloping valley on the flank of a glacial moraine. The site received concentrated runoff from street drains along Harvard Road, a long, steep roadway ending in a cul-de-sac located above the valley.

SITE CONDITIONS AND PROBLEMS: A large gully had formed over time as the result of the concentrated discharge of storm water into the Arboretum from street drains. The gully which, passed close to a residence, threatened to undermine the home by a process of down cutting and channel enlargement. In addition scour and channel erosion in the gully during major storms resulted in sediment deposition and the formation and encroachment of a sediment delta at its terminus in a grassy, meadow. The soils are primarily fine sands and sandy loams that are susceptible to surficial erosion.

TREATMENT OBJECTIVES AND CONSIDERATIONS: Treatment strategy basically consisted of reducing velocity and energy of runoff entering the Arboretum from street drains along Harvard Road. A reduction in velocity and tractive stress of water flow in channels can be accomplished by either spreading the flow out (decreasing depth of flow) or lengthening the channel (decreasing gradient). The latter can be achieved by introducing channel sinuosity (adding meander bends). Care must taken when adding sinuosity to insure the meander bends are not too

W. Goldsmith et al., *Bioengineering Case Studies: Sustainable Stream Bank and Slope Stabilization*,
DOI 10.1007/978-1-4614-7996-3_7, © Springer Science+Business Media New York 2014

tight because tractive stress tends to increase with channel curvature. Energy dissipation can be achieved by adding stepped-pools along the channel that are armored or rocked. Increased resistance to scour and downcutting erosion can be provided by lining the channel with rocks and/or erosion control netting. Surficial erosion control around and next to the channel (or interfluve areas) was also required. This can be provided by seeding and use of erosion control blankets in combination with live staking. The site is highly visible to and frequented by a large number of visitors to the Arboretum; it was important therefore to use stabilization measures and techniques that are attractive and environmentally sensitive.

TREATMENTS SELECTED:

- **Rocked pools.** A series of rocked, "stepped-pools" (low-height boulder drops) were constructed at critical locations along the length of an artificial sinuous channel (see Fig. 1). These pools dissipated the energy of runoff flowing along the channel. A grated, flared entrance structure was placed at the head of the rocked pools where concentrated runoff entered the Arboretum from street drains along Harvard Road (see Fig. 2).

- **Sinuous channel.** An artificial channel with a sinuous (curved) planform was constructed to lengthen the flow path, decrease the flow gradient, and to convey concentrated runoff from the street drains at a non-eroding velocity. The channel was lined with a natural coir geotextile and coarse gravel or small cobble-size rocks to increase resistance to tractive stresses (see Fig. 3).

Fig. 2 Grated, flared entrance structure and rocked pool

Fig. 3 Construction of meandering channel and rocked pools

- **Erosion control blanket.** An erosion control blanket was placed on the interfluve areas and along the edges of the sinuous channel. The blanket was anchored to the ground using staples and pegs. Topsoil was spread on the ground followed by a seed mix of native grasses and plants before placing the erosion control blanket.

- **Live staking.** Live staking was used to reestablish vegetation and help stabilize interfluve areas around the sinuous channel. The live stakes, mostly willow willow (*Salix spp.*), were cut in Nichols Arboretum or nearby natural areas along the Huron River.

OBSTACLES TO IMPLEMENTATION: There were no significant obstacles. All parties (the City of Ann Arbor, University of Michigan, and residents at the bottom of Harvard Road) were anxious to solve the problems noted previously in the most cost-effective and environmentally sensitive manner possible.

PERFORMANCE EVALUATION: The system has performed well during and following intense rainstorms (see Fig. 4). The combination of sinuous channel and rock armored, stepped pools have decreased flow velocity and dissipated energy sufficiently so that no channel erosion or gullying has occurred. Flow has been restricted to the sinuous channel; meander bend cutoffs have not materialized. The area is now well vegetated and visually attractive.

Fig. 4 Completed project (1-year later) during major storm event

BENEFITS AND LESSONS LEARNED:

1. The gully repair and watershed restoration were accomplished in an environmentally sensitive manner by taking what had previously been a liability, viz., an erosive discharge of concentrated runoff and converting it into an asset, viz., a controlled, low-velocity flow that now provides a source of water to a grassy meadow area.

2. This project demonstrates that gully repair can be accomplished without the need for large structures (check dams) and cement-lined channels.

3. A sinuous or meandering channel is a reasonable alternative to a level spreader as a means of decreasing erosive tractive stresses.

ADDITIONAL READING:

OHM Engineering (2009). Harvard Drain and Nichols Arboretum Improvements, Construction Drawings and Notes, June 23, 2009.

Schor, H. and Gray, D.H. (2007). Landforming: An Environmental Approach to Hillside Development, Mine Reclamation and Watershed Restoration John Wiley & Sons, New York: NY, August 2007.

Gray, D.H. and Sotir, R. (1996). Biotechnical and Soil Bioengineering Slope Stabilization. John Wiley & Sons, New York: NY

PROJECT #7: MALLETTS CREEK

PROJECT TYPE: **Stream bank repair and stabilization**

PROJECT SCALE: Large

CLIENT/OWNER: Washtenaw Co. Water Resources Commissioners Office and City of Ann Arbor

TECHNIQUES EMPLOYED: **Rock vanes, Coir logs, Channel profile mod ECBs and live staking**

GEOGRAPHIC LOCATION: Southeastern Ann Arbor, MI

Fig. 1 Malletts Creek bank protection

GEOMORPHIC SETTING: Malletts Creek watershed is 11-square miles in area; the creek flows into the Huron River. The watershed lies primarily in an urban area. During the last 40 years, the watershed has been extensively developed with shopping malls, new subdivisions, parking lots, etc. About 37 % of the land is covered with impervious surfaces.

SITE CONDITIONS AND PROBLEMS: Malletts Creek had been degraded over time as a result of water quality and flow capacity problems. Major water quality problems included high phosphorus loading and sedimentation. Water quantity problems were caused by high flow velocities that resulted in both bank and channel erosion (see Fig. 2). Degraded streambeds and frequent/severe peak flows diminished aquatic habitat. The soils are primarily fine sands and sandy loams that are susceptible to scour erosion.

TREATMENT OBJECTIVES AND CONSIDERATIONS: Multiple restoration objectives were developed for both the creek and its watershed. A public information component was included to inform and involve citizens about proposed restoration

W. Goldsmith et al., *Bioengineering Case Studies: Sustainable Stream Bank and Slope Stabilization*,
DOI 10.1007/978-1-4614-7996-3_8, © Springer Science+Business Media New York 2014

Fig. 2 Eroded channel and stream bank

activities. The main treatment goals were to address needed structural repairs (e.g., broken culverts and cracked outlet structures), to reduce phosphorus loading by 50 %, to increase habitat for fish and wildlife, to control stream velocities, and to stabilize stream banks. Well-publicized public meetings were held with stakeholders, and frequent interactions occurred during construction with homeowners.

A total length of 8,700 linear feet of stream channel and bank required treatment. Ancillary measures included the construction of detention basins to slow down polluted storm water runoff from adjacent developed areas in the watershed. A detailed computer modeling study was conducted to predict flow rates and flood elevations as well as phosphorus levels in the creek. The total estimated cost of all restoration activities in the watershed, including stream channel and bank stabilization, was about $19.1 million. The estimated cost of stream bank stabilization was $2.1 million (11 % of total). The largest cost components were initial structural repairs to streams ($4.8 million) and detention pond improvements ($4.4 million).

TREATMENTS SELECTED:

Both conventional and bio-stabilization measures were selected for stream-bank stabilization. All stream channel/bank stabilization was preceded by extensive removal of invasive woody vegetation accompanied by preparatory grading work. Rock armor (riprap) was mainly used along high-velocity reaches (>5 ft/s) such as outside bends while coir logs were placed along other sections. These measures were used to provide protection against scour erosion at the toe. Rock vanes were often placed along tight outside bends to direct flow away from the bank. The channel topography was modified to provide a wider flow path along the channel bottom; meanwhile the banks were graded back to a 2:1 (H:V) slope and covered with a biodegradable straw or coconut fiber ECB. Live willow stakes were inserted through the ECB to provide additional bank protection. Where space allowed the stream bank was "stepped" to provide a shallow terrace or bench above the rock toe to accommodate and slow down intermediate flood flows.

- **Rock armor.** Conventional rock armor (riprap) was widely used on this project as toe protection (refer to Figs. 1 and 3). Rock armor was used in higher velocity reaches and along outside bends. There was no attempt to purposely vegetate the rock with live cuttings using either the "joint planting" or "bent willow pole" method. On the other hand, the rock selected was relatively rounded and attractive in appearance. In addition, larger rocks were placed at the bottom and inset into the channel to forestall displacement and undermining during high flow events.

- **Coir logs.** Coir logs consisting of coir fiber rolls approximately 1-ft in diameter were staked in place at the toe of the stream bank. These were placed either singly or stacked vertically in a double row. Coir logs were used as an alternative to rock armor in lower velocity reaches of the channel. Coir log can be vegetated by inserting live stakes or cuttings; they also lend themselves to natural revegation by volunteer plants. Coir logs were occasionally protected with rock as well (refer to Figs. 3 and 4).

Fig. 3 Toe protection—riprap and coir logs

Fig. 4 Coir log toe protection

- **Channel profile modifications.** The channel profile was modified in both the longitudinal and cross-channel directions to improve stream function (see Fig. 5). The channel width was widened along its bottom to accommodate increased flow; the sides were also graded back to a more stable 2:1 (H:V) angle. Rock was placed along the bottom of steeper reaches in a manner that resembled "boulder cascades" to limit excessive bed degradation.

Fig. 5 Modified bank with terrace

- **Rock vanes and cross vanes.** Rock vanes, elongated rock prisms, were placed at strategic locations to deflect or redirect impinging flow away from the bank (see Fig. 6). The vanes were oriented in an upstream direction at roughly a 30° angle with the bank. Cross vanes serve a similar purpose but extend from both banks and V-upstream. They deflect the current into the center of the stream.

- **ECBs and live staking.** Both straw and coconut fiber erosion control blankets (ECBs) were used to protect stream banks above either the rock armor or coir logs placed at the toe. Live stakes cut largely from phreatic species such as willow (*Salix spp.*) were inserted through the ECBs to provide additional protection and to help reestablish vegetation on the stream banks (see Fig. 7)

Fig. 6 Rock vanes

Fig. 7 ECB's and live staking

Fig. 8 – Sediment accumulation above cross- vane

OBSTACLES TO IMPLEMENTATION: The main obstacle to implementation of the stream bank stabilization work was initial concerns by residents near the creek about tree removal, and how this removal and attendant grading might affect their views and privacy. These concerns were allayed by reassurances that the loss of vegetation was only temporary and that the stream bank work was an important component of the overall watershed improvement goals.

PERFORMANCE EVALUATION: The system performed well during and immediately after intense rainstorms shortly after construction in March 2012. This was a severe test for a stream bank protection system which can be vulnerable at this juncture. The live cuttings and seedings (under the ECBs) did not have sufficient time by then to grow and establish fully. Channel degradation that appears to have been arrested and considerably less sediment is being flushed through the creek to the Huron River. The rock vanes provide a superior and more diverse aquatic habitat; in addition, sediment has started to collect behind the vanes as well (see Fig. 8).

BENEFITS AND LESSONS LEARNED:

1. Stream bank and channel stabilization were important and critical components of the Malletts Creek watershed restoration program.

2. The combination of conventional measures (rock armor) and bio-stabilization techniques (coir logs, ECBs, and live staking) have functioned well together.

3. Initial concerns by residents adjoining the creek were allayed by reassurances that loss of vegetation was only temporary and that the stream bank work was an important component of overall watershed improvement goals.

ADDITIONAL READING:

Washtenaw County (2011). Mattletts Creek Restoration Project: Final Report, available on line at: http://www.ewashtenaw.org/government/ drain_commissioner/dc_webWaterQuality/malletts_creek/dc_mc_mcrp.html

NCHRP (2005). Environmentally Sensitive Channel and Streambank Protection Measures, NCHRP Report 544. Guidance manual prepared for the Natl. Coop. Hwy. Research Program, Washington, DC.

Gray, D.H. and Sotir, R. (1996). Biotechnical and Soil Bioengineering Slope Stabilization. John Wiley & Sons, New York: NY

PROJECT #8: TOBOGGAN HILL

Fig. 1 View of Toboggan Hill

PROJECT TYPE: **Erosion control**

PROJECT SCALE: Small

CLIENT/OWNER: Wayne County Parks

TECHNIQUES EMPLOYED: **Live fascines, Live staking, RECPs, Re-vegetation**

GEOGRAPHIC LOCATION: Edward Hines Drive Garden City, MI

GEOMORPHIC SETTING: Toboggan Hill is artificial mound (hill) consisting of industrial waste covered with a clay cap and top soil layer (Fig. 1). Toboggan runs were constructed on one side of the hill. The facility was incorporated into a county park running along Edward Hines Drive and paralleling the Rouge River.

SITE CONDITIONS AND PROBLEMS: The mounded waste was covered with a clay cap and top soil layer. A grass cover was established on the south side of the mound and the site converted to a toboggan hill. The slope gradient on the south facing side of the landfill mound is relatively gentle—approximately 3:1 (H:V) on average. The toboggan runs are located on this side. The slope gradient on the opposite side, on the other hand, is quite steep—2:1 (H:V) or higher—and prone to serious erosion. The slope on this steep side abuts a major highway that constrains any regrading or profile adjustment (see Fig. 2).

TREATMENT OBJECTIVES AND CONSIDERATIONS: A cantilever retaining wall was constructed at the base of the north facing slope along the roadway. This wall buttresses and protects the toe of the slope; however, the grade above still remains quite steep, much steeper than the limits normally imposed for construction and revegetation of landfill covers. An erosion control blanket (ECB) or turf reinforcement mat (TRM) was considered for use on the steep 2:1 north facing slope. There was concern this treatment would not resist both surficial erosion and the threat of shallow surface sliding of the soil layer atop the clay cap. The use of an erosion control netting in combination with life fascines was selected to meet this challenge.

W. Goldsmith et al., *Bioengineering Case Studies: Sustainable Stream Bank and Slope Stabilization*,
DOI 10.1007/978-1-4614-7996-3_9, © Springer Science+Business Media New York 2014

Fig. 2 Steep landfill slope abutting highway

TREATMENTS SELECTED:

- **Live fascines.** A coconut fiber netting was placed on the slope surface and also run into shallow trenches dug at periodic intervals on contour. Live fascines were placed in the trenches and staked in place over the netting using inert construction stakes (see Fig. 3). The fascines firmly anchored the netting to the slope and prevented rills or small gullies from developing. The sinker roots from the fascines bundles also tended to prevent shallow sliding by penetrating partially through and anchoring the soil layer to the underlying clay cap.

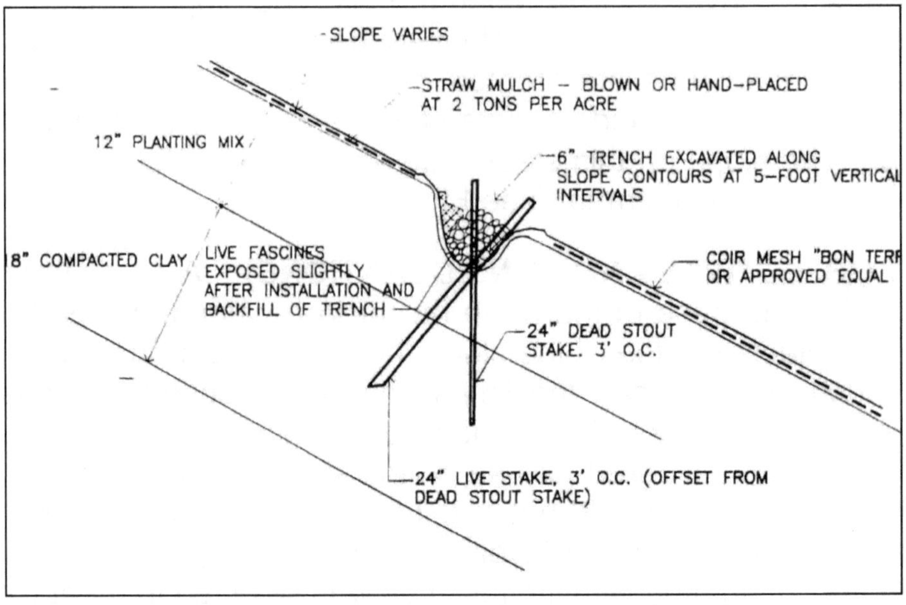

Fig. 3 Schematic drawing of live fascine and netting system

Fig. 4 – Live fascine anchored with live and inert stakes

- **Live stakes** were also used in addition to inert construction stakes to secure the fascine bundles in the trenches (see Fig. 4). The live stakes rooted and provided additional anchorage and erosion control.

- **Transplanting.** Fountain grass, Tahitian lilies, and Meiland roses were planted on the slope as well to help prevent erosion and for landscaping.

OBSTACLES TO IMPLEMENTATION: Some concern was voiced by Parks personnel about establishing woody plants (willows) atop a landfill. However, no evidence was presented that the roots would penetrate through the clay cap into the industrial waste below. Plant roots tend to avoid heavily compacted soils or layers with little or no oxygen and water. Furthermore, woody plants (trees) have been purposely planted atop much larger, nearby municipal waste landfills in efforts to convert them to post-closure recreational purposes (e.g., South Lyon Township Community Park)—without apparent adverse effects.

PERFORMANCE EVALUATION: The site has remained stable since construction; the soil bioengineering installation has developed into a dense vegetative cover (see Fig. 5). There is no evidence of any erosion on the steep, north facing slope of the landfill cover. The live fascines and the RECP (coir netting) have worked together in a mutually complementary manner. The fascines break the slope up into short segments and prevent rills from extending beyond the buried RECP in the fascine trenches. The question of whether the roots from the willow bundles have penetrated through the top soil and clay cap into the underlying waste remains somewhat unresolved; no exhumation tests have been performed to date.

Fig. 5 Appearance of project site after 2 years—June 2001

LESSONS LEARNED:

1. The combined use of live fascines and a RECP provides excellent resistance to both surficial erosion and shallow mass wasting (face sloughing).

2. Skeptics have voiced concern about this approach to landfill cover stabilization. The concern focuses on the possibility of root penetration from fascines through the clay cap into the underlying waste. This concern could be addressed and resolved by excavating a test area of the treated cover to determine the extent (if any) of root penetration into the underlying waste.

ADDITIONAL READING:

Gray, D.H. (2001). Combined Use/Installation of Rolled Erosion Control Products with Soil Bioengineering Measures. *Proceedings*, 15th GSI/GRI Conference, Houston, TX

Gray, D.H. and Koerner, R.M. (2002). Post Closure Beneficial Uses of Landfills. *Proceedings*, 16th GSI/GRI Conference, Philadelphia, PA

PROJECT #9: ARGO CASCADES

PROJECT TYPE: **Dam bypass watercourse**

PROJECT SCALE: Medium

CLIENT/OWNER: City of Ann Arbor

TECHNIQUES EMPLOYED: **Stepped pools, Boulder spurs, ECBs and hydroseeding**

GEOGRAPHIC LOCATION: Adjacent to Huron R. below Argo Dam Ann Arbor, MI

(reprinted w/ permission from the City of Ann Arbor)

Fig. 1 Argo Cascades canoe passage

GEOMORPHIC SETTING: Argo Cascades (Fig. 1) is a former millrace or canal, about 1,500 ft long below Argo Dam, that supplied water to a Detroit Edison hydropower station. The canal was held in place by an earthen embankment that was constructed parallel to and adjacent the Huron River. The site lies in a valley that cuts through glacial moraines on either side. Canoes and boaters using the Huron River could enter the canal from an impoundment above Argo Dam; however, a portage from the canal was required to reenter the river below. The millrace canal was transformed into a series of pools connected by short drops that have eliminated the need for a portage.

SITE CONDITIONS AND PROBLEMS: The Argo Cascades replaced a previous millrace canal that had tended to become stagnant and unsightly over the years. Questions were also raised about the stability and safety of the earthen embankment that held the millrace in place. Furthermore, the end of the millrace often became congested with canoes and kayaks faced with dragging their boats out of the water and carrying them back down to the river.

W. Goldsmith et al., *Bioengineering Case Studies: Sustainable Stream Bank and Slope Stabilization*,
DOI 10.1007/978-1-4614-7996-3_10, © Springer Science+Business Media New York 2014

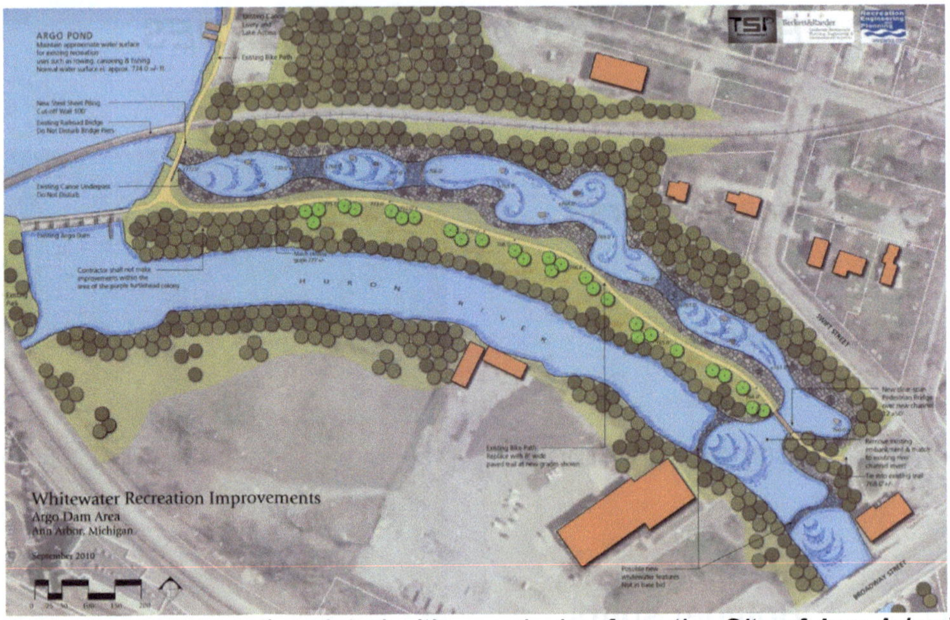

(reprinted with permission from the City of Ann Arbor)

Fig. 2 – Schematic of connecting-pool bypass

TREATMENT OBJECTIVES AND CONSIDERATIONS: Construction of the bypass addressed several issues and objectives, namely, (1) eliminated the need for a portage around Argo Dam, (2) greatly reduced the threat and consequences of a possible failure of the earthen embankment, and (3) provided an attractive watercourse with some whitewater features for boaters to enjoy. A system of pools connected by short drops or passages was adopted for this purpose (see Fig. 2). This bypass would take boaters from the impoundment behind the dam back down to the river below. Enough pools and connecting passages were required to gradually transfer boats in gentle increments over a total elevation change of 14 ft. Other important considerations were the design and materials used in constructing the artificial watercourse to make it appear like a natural feature. This dictated the use of rock spurs and boulder clusters at the passages and minimal use of exposed concrete.

TREATMENTS SELECTED:

- **Connecting pools:** The former millrace was drained (see Fig. 3), excavated, and graded to accommodate a total of 8 pools placed at gradually descending elevations. When full of water each pool's surface is separated by a vertical elevation difference or drop of approximately 2½ ft. Natural boulders were used to form and protect the outside edge of the pools. The appearance of the first pool during construction is shown in Fig. 4.

Fig. 3 Millrace after draining and prior to grading

Fig. 4 First pool during construction

- **Boulder spurs.** The connecting passages (drops) were formed by boulder spurs and clusters that extended out from the bank on either side towards the middle of the watercourse. An opening was left between the two to allow flow of water and transfer of boats (canoes and kayaks) from the upper to adjacent lower pool (see Fig. 5).

(reprinted with permission from the City of Ann Arbor)

Fig. 5 Boulder spurs extending from banks of bypass. The openings between the spurs formed the connections or passages between pools

- **ECBs and hydroseeding.** The newly graded and exposed sides of the bypass were protected by ECBs that were hydroseeded with grass. The sides were graded at a relatively gentle 3:1 (H:V) slope to minimize surficial erosion. Native, volunteer plants will eventually invade and provide additional landscaping and erosion control.

OBSTACLES TO IMPLEMENTATION: The only real obstacle to implementation was some initial concern about the project cost ($1.1 million). The project was also held up to some extent because it was embroiled in a larger controversy and because of proposal to remove Argo Dam. If the dam had been removed, this would have eliminated the need for a bypass. Once the decision was made by the community to keep the dam, plans proceeded quickly, and with substantial citizen support to construct the bypass, concerns about cost were outweighed by the recreational advantages and amenities that the bypass would provide.

PERFORMANCE EVALUATION: The Argo Cascades has become a magnet for boaters looking for a little excitement without having to worry about a portage. A trail along the bypass now provides an important link in a trail system designed to traverse the length of county. Visitors can now observe stunning views of the bypass and its connecting pools (see Fig. 6) from the trail. The passages between pools have required a certain amount of testing and adjustment

Fig. 6 View of Argo Cascades connecting pools looking upstream

to ensure safe passage. At present only kayaks are allowed to run through all the connecting pools. Canoes must put in at the final pool at the bottom. Water flow into the Cascades from the pond upstream of Argo Dam is shut off and boat traffic prohibited during low flow in the river to allow a sufficient amount of water to pass through the dam control gates.

BENEFITS AND LESSONS LEARNED:

1. Bioengineering measures can be used effectively to construct a dam bypass system consisting of connecting pools using naturally occurring rocks and boulders.

2. A considerable amount of post-construction adjustment and "tweaking" may be required to make the connections between the pools navigable.

ADDITIONAL READING:

NCHRP (2005). Environmentally Sensitive Channel and Streambank Protection Measures, NCHRP Report 544. Guidance manual prepared for the Natl. Coop. Hwy. Research Program, Washington, DC.

Kayaking Through the Argo Cascades. A You-Tube Video, June 17, 2012 http://www.youtube.com/watch?v=nWn4NFl4u5U&list=UU0tji3zKiGNBEDM8yKHjA6A&index=1

Western USA

PROJECT #10: ASAAYI LAKE

PROJECT TYPE: Watershed repair and restoration

PROJECT SCALE: Large

CLIENT/OWNER: The Navajo Nation EPA Water Quality Division

TECHNIQUES EMPLOYED: **Landform grading, rock ramps, re-vegetation**

GEOGRAPHIC LOCATION: Chuska Mountains, Navajo Reservation New Mexico (*Asaayi Lake also known as Bowl Lake Lake on USGS maps*)

Photo courtesy of H.J. Schor

Fig. 1 Deep gully formed in valley

GEOMORPHIC SETTING: A narrow valley over 1,400 ft long with a tributary drainage area to Lake Asaayi of 200 acres. The original, natural grade in the valley was 6 %. The total runoff flowing down the valley was approximately 800 cfs (based on a 25-year storm condition).

SITE CONDITIONS AND PROBLEMS: Deeply incised gullies (Fig. 1) discharged sediment into Asaayi Lake that threatened to destroy its recreational value. The main degradation or incision occurred along the center of the valley. Deep erosion channels also developed on the west side from runoff entering the valley from adjacent mountains (Fig. 2). Much of the erosion was also caused by the construction of a logging road which later became the primary access road to the lake and its recreational facilities. Uncontrolled seasonal grazing with its attendant trampling and destruction of ground cover also exacerbated erosion problems.

TREATMENT OBJECTIVES AND CONSIDERATIONS: The main objective was to "heal" the damaged valley by reshaping its landform via geomorphic modification. This alteration replicated the South Western USA "wash" landscape which is less prone to erosion and consequently to the danger of lake siltation. The original plan

DOI 10.1007/978-1-4614-7996-3_11,

Photo courtesy of H.J. Schor

Fig. 2 Deep, channelized tributary gully

that was considered by the NRCS called for construction of a series of large check dams to intercept runoff, reduce flow velocities, and trap sediment. This plan was rejected by the Navajo EPA and the Navajo Tribal Council.

These agencies determined that the plan was too invasive and too damaging to the natural beauty of the forested mountain terrain. In addition it would have left the valley in its eroded state. This outcome was not acceptable to the tribe from either an esthetic and recreational use standpoint nor did it allow any future use of the valley, e.g., for seasonal grazing.

TREATMENTS SELECTED:

- **Geomorphic modification and regrading.** The main axis of the valley was regraded to remove erosional scars and to create a broad, natural analog wash. This topographic alteration included the following elements:

    1. The central valley was regraded to a 100-ft width at its upstream head and expanded to a 275-ft width at its downstream terminus. The reconfigured wash depth ranged from 10 to 20 ft. Its side banks were laid back to slopes with gradients ranging from 1.5:1 to 3:1 (H:V).
    2. Three of the tributary side channels were regraded starting at their upper ends into broader base washes and flared out in the direction of the flow at the downstream outlet or confluence with the main channel (see Fig. 3).

- **Rock ramps.** Three grade reduction/erosion control structures (rock "ramps") were installed on the main channel to reduce flow gradients to between 1 % and 4 % between reaches (see Fig. 4). One ramp was placed at the head scarp upstream, one at midpoint and one at the downstream discharge point.

Photo courtesy of H.J. Schor

Fig. 3 Regraded confluence between tributary and main valley

Photo courtesy of H.J. Schor

Fig. 4 Rock ramp construction in main wash

- **Re-vegetation.** Existing trees or shrubs in areas to be graded were preserved where possible by conducting grading around them. Revegetation was accomplished by hydro-mulching of all graded areas with native seed mixes consisting primarily of Indian Rice Grass and Western Wheat Grass (see Fig. 5). The site was fenced to prevent the premature entry of livestock prior to establishment of a durable vegetative groundcover.

Photo courtesy of H.J. Schor

Fig. 5 Revegetation of valley with native grasses

OBSTACLES TO IMPLEMENTATION: One of the main obstacle to implementation of erosion repair and control was cost. This limitation required scaling back on the number of rock-ramp, grade reduction structures that were used. A diagonal berm was placed across the mouth of one of the tributary channels where it discharged into the main channel in lieu of a ramp. The overall design that was ultimately adopted and constructed was considerably less expensive, however, than the original $1.5 million cost estimate for installation of check dams in the valley.

PERFORMANCE EVALUATION: A local private contractor was selected who completed the construction in 3 months. Straw was applied on top of the hydro-mulching to provide additional interim protection prior to ground cover establishment. Lack of precipitation and later high winds blew much of the straw cover away preventing early germination. In spite of these setbacks, which also included funding reductions, relaxed specifications, and unfavorable weather conditions, the project fared exceptionally well.

The most important goal of all, namely eliminating the destructive erosion in the valley while maintaining the natural character of the region was achieved. The erosion scars have healed, siltation of the lake has stopped, and the valley is beginning to blend naturally into the surrounding landscape (see Fig. 6). After the first year only minor gullying in the channel bottom and side slopes had occurred. This was to be expected because a substantial vegetative cover had not yet become established, and residual gradients on the project of up to 3 % were still in the erosive range.

Photo courtesy of H.J. Schor

Fig. 6 Appearance of main valley 3 years after treatment

BENEFITS AND LESSONS LEARNED:

1. This project is a good example of how landform grading (aka geomorphic modification) and revegetation lend themselves well to repairing a damaged landscape.

2. Repair of serious gully erosion and sedimentation problems were accomplished without the need for constructing costly check dams.

3. The treatments selected were both esthetically pleasing and environmentally compatible and acceptable to the Navajo Tribal Council.

4. Treatment has successfully eliminated erosion scars (gullies) in the valley and restored the original Southwestern USA "wash" landscape.

ACKNOWLEDGMENT

Helpful information about this project was provided by Horst Schor, of H.J. Schor Consulting, who developed the Landform/Geomorphic restoration concept, prepared preliminary design plans, and supervised construction. His contribution is greatly appreciated and acknowledged.

This case study is derived from material appearing in the book **Landforming** coauthored by Schor and Gray, published by John Wiley & Sons, 2007. This derivative material is reproduced with permission of John Wiley & Sons, Inc.

ADDITIONAL READING:

USDA (2002). Asaayi Lake Northwest Drainage Channel Shaping and Grade Stabilization. Design documents prepared for the Navajo Nation and Fort Defiance SWCD. Natural Resources Conservation Svc., March 2002

Schor, H.J. and Gray, D.H. (2007). Landforming: An Environmental Approach to Hillside Development, Watershed Restoration, and Mine Reclamation. John Wiley and Sons, Hoboken, NJ.

Schor, H.J. (2008). Restoring a Southwestern Valley, *Erosion Control*, Vol. 15, No. 6,

PROJECT #11: HOLLYWOOD HILLS

PROJECT TYPE: **Mass grading and canyon fills**

PROJECT SCALE: Large

CLIENT/OWNER: City of Los Angeles DWP Hollywood Hills Master Association

TECHNIQUES EMPLOYED: **Landform grading, drainage swales, re-vegetation**

GEOGRAPHIC LOCATION: Hills and canyons surrounding Lake Hollywood, Los Angeles, CA

Photo courtesy of H.J. Schor

Fig. 1 Aerial view of site

GEOMORPHIC SETTING: Rugged terrain surrounding Lake Hollywood consisting of larger, lower 11-acre water storage reservoir and upper, much smaller one (see Fig. 1) in the Hollywood Hills [below the famous landmark Hollywood sign]. Total site acreage including watershed area is some 200 acres. The fill disposal, facilities locations and mitigation sites covered about 40 acres. Only one of the canyon fills, Area "F," is described herein (see Fig. 2).

SITE CONDITIONS AND PROBLEMS: The excavation of soil from large underground chambers (water tanks) and rock materials from tunneling operations generated over 1.0 million cubic yards of waste. This spoil was disposed of in tributary canyons around the lake. Environmental mitigation had to be performed to counter the taking of coastal sage and riparian wetlands and blue-line stream beds.

TREATMENT OBJECTIVES AND CONSIDERATIONS: The overburden from the excavations and tunneling operations had to be disposed of in an environmentally responsible manner. Excessive costs and transportation constraints precluded offsite disposal. Accordingly, a decision was made to place the spoil in tributary canyons around the lake. However, this type of disposal had to be done in a way that was also acceptable to surrounding homeowners in the area. Ultimately, a plan was agreed upon by all parties that would allow placement of the overburden in selected canyons that met the following main conditions:

W. Goldsmith et al., *Bioengineering Case Studies: Sustainable Stream Bank and Slope Stabilization*,
DOI 10.1007/978-1-4614-7996-3_12,

Photo courtesy of H.J. Schor

Fig. 2 Overhead view of project site showing different fill areas

1. Suitability from the standpoint of accessibility for the grading equipment to prevent scarring of adjacent hillside areas

2. Ability to minimize alteration of existing topography and impact on views over the watershed from surrounding residences.

3. Sufficient capacity and ability to shape the fills into natural landforms that would harmonize with the adjacent topography.

4. Ability to reestablish a natural hydrological pattern that would supply moisture to the native vegetation.

These conditions required a canyon fill plan that resulted in essentially replicating the original canyon topography as far as possible.

TREATMENTS SELECTED:

- **Geomorphic modification.** Fill placement in a canyon using landform grading principles. The creation of concave slope forms would collect, concentrate, and deliver runoff to locations where it could sustain native plant communities. This overburden placement in the canyons shared the following design attributes and characteristics:

  (a) The fills were placed and shaped to replicate the original valley or canyon topography but at higher overall elevation (see Fig. 3).

Fig. 3 Replication of original canyon surface topography at higher elevation

Fig. 4 Use of curvilinear slope forms in canyon fill

(b) Side slopes of valley fills had concave and convex slope forms incorporated either to simulate the underlying original topography or mimic side slope forms typical of the area (see Fig. 4).

(c) Linear ridge features were included as well as curvilinear- and elbow-shaped swales.

- **Drainage swales.** All valley and slope forms replicated the original slope hydrology so as to maintain runoff patterns and support original plant distributions. Traditional concrete surface slope drainage structures were avoided to allow both runoff and infiltration to sustain plant life instead of filling storm drain pipes. Where possible rock lined, curvilinear ditches with porous bottoms were placed in lieu of concrete linings.

- **Re-vegetation.** Vegetation was selected and placed so that it was compatible with slope hydro-geology. Woody shrubs and trees were placed in wetter swales and valley bottoms and grasses and forbs in drier interfluve areas.

OBSTACLES TO IMPLEMENTATION: One of the main obstacles to implementation was lack of familiarity on the part of contractors and field personnel with landform grading practices. For example, during placement of the first canyon fill, the resultant shapes did not appear natural and instead began to resemble the old conventional grading shapes, including an earthen dam structure at the foot of the canyon. It became clear that further on-site training on the use of landform grading designs and techniques would be required along with closer supervision of ongoing operations. Dozers and scrapers were used to remove the "dam" structure and build up the side slopes while simultaneously raising the valley floor per original intent. The remainder of the fill was constructed as planned. As time went on a certain amount of pride was exhibited by all personnel involved as they became accustomed to landforming techniques and as they realized that a landformed fill really did "look" better than fills constructed in the conventional manner.

PERFORMANCE EVALUATION: There has been minimal impact on the quality of life, visual esthetics, and wildlife as a result of the canyon fill placement. Area "F" proved that after five years this area is hardly distinguishable to the untrained eye from the original canyon scenery in terms of both landform and vegetative patterns (see Fig. 5). The Department (DPW) was pleased with the results. Landform Grading and Revegetation allowed them to gain community support and proceed with their project to meet state mandated deadlines. This approach, furthermore, enabled them to perform all required environmental mitigation on-site thus saving the cost of property acquisition elsewhere in addition to avoiding a logistically more difficult off-site disposal installation and management.

Fig. 5 Appearance of Fill Site "F" after 2 years

BENEFITS AND LESSONS LEARNED:

1. The introduction of Landform Grading concepts eliminated citizen objections. Unanimous acceptance by the Community was ultimately achieved following protracted battles and disagreements over conventional grading proposals that were initially proposed.

2. Expensive offsite mitigation requirements were eliminated. The necessary work was accomplished within the actual construction site and consequently did not involve any expensive and distant offsite property acquisition and facilitated the long-term maintenance responsibilities.

3. One unplanned and unexpected problem was encountered during the construction process that should be kept in mind in future projects. The material excavated from the tunnel was shaly and devoid of any organic compounds. This material was placed atop the surface of one of the canyon fills instead of being placed deep within the fill. Transplants initially did not do well and seeds placed on the fill surface failed to germinate in spite of repeated applications of hydromulch, fertilizer, and regular irrigation.

ACKNOWLEDGMENT

Helpful information about this project was provided by Horst Schor, of H.J. Schor Consulting, who developed the Landform/Geomorphic restoration concept, prepared preliminary design plans, and supervised construction. His contribution is greatly appreciated and acknowledged.

This case study is derived from material appearing in the book **Landforming** coauthored by Schor and Gray, published by John Wiley & Sons, 2007. This derivative material is reproduced with permission of John Wiley & Sons, Inc.

ADDITIONAL READING:

Schor, H.J. and Gray, D.H. (2007). Landforming: An Environmental Approach to Hillside Development, Watershed Restoration, and Mine Reclamation. John Wiley and Sons, Hoboken, NJ.

PROJECT #12: GEYSERVILLE

PROJECT TYPE: **Stream bank and channel stabilization**

PROJECT SCALE: Large

CLIENT/OWNER: California Department of Transportation

TECHNIQUES EMPLOYED: **Rock vanes, LSTP, Live siltation, Pole planting**

GEOGRAPHIC LOCATION: Bridge crossing on SR 129 near Geyserville, CA

Fig. 1 Stream bank protection site

GEOMORPHIC SETTING: Bridge site is located on the right descending bank of the Russian River, a broad meandering river that flows on deep alluvial soils (Fig. 1). Flows in the river vary seasonally, attaining values as high as 10,000 cfs during winter floods. The soils at the site consist of riverbed alluvium and floodplain deposits. These soils are well drained and not subject to ponding. They are mainly cohesionless and vulnerable to erosion and scour.

SITE CONDITIONS AND PROBLEMS: Over time, high velocity flows along an outside bend of the river upstream of the bridge abutment had undermined the bank and created a sharp, "hook" like bend (see Fig. 2) that threatened to cut behind the abutment and isolate the bridge. Reacting to this threat Caltrans placed tons of large riprap (RSP) during a storm in January 2010 under near bankfull conditions, as an emergency measure to protect the bridge abutment. However, the river continued to migrate west and eroded the adjacent land and vineyard. The bank erosion and subsequent migration shaped the new bankline into a severe crescent-shaped "hook" which tended to exacerbate the impingement of flows and direct the passing flows parallel to the bridge and towards into the opposite bank.

W. Goldsmith et al., *Bioengineering Case Studies: Sustainable Stream Bank and Slope Stabilization*,
DOI 10.1007/978-1-4614-7996-3_13,

Fig. 2 Initial conditions showing extent of bank retreat and erosion

Ideally flood flows should proceed through the bridge in as direct a manner as possible. So the "hook" represented a topographic anomaly, potentially as damaging as the lateral migration. The top of the bank (TOB) is essentially flat and is about 6–9 ft (1.8–2.7 m) above the level of the river under normal low flow conditions. According to a USDA soil survey map the reach of riverbank at the bridge site is classified as very gravelly sandy loam, 0–2 % slope.

TREATMENT OBJECTIVES AND CONSIDERATIONS: Prevention of continued erosion/scour at the toe of the riverbank next to the bridge abutment required the use of either bank armor or redirective measures such as rock vanes or bendway weirs. Emergency planning meetings were held between Caltrans and various resource agencies to address the bank erosion threat. The original plan was to adopt an aggressive design that included skimming of built up bar material opposite the scoured bank, construction of an earthen levee, and use of rock armor. This design is quite common historically under similar circumstances. The resource agencies balked at the design due to the scale of the fix, impacts to the river, and lack of restored habitat. A contingency plan was also developed in conjunction with the original plan. More emphasis was placed on the use of a redirective approach to realign higher scour velocities away from the bank. This is a more environmentally friendly approach because it displaces flow away from the bank towards the center of the stream, does not require laying back the bank, provides greater flow diversity, and promotes sediment deposition.

TREATMENTS SELECTED:

A combination of treatments and repair measures that were compatible with site conditions and environmental constraints were selected to meet the stabilization requirements and goals. A series of redirective and bioengineering methods were employed in the Fall of 2010 along the right descending bank in expectation of high flows that would occur in the Winter of 2010 and 2011. The primary channel and bank protective methods that were chosen were the following:

- **Rock vanes.** Elongated, rock prisms with a triangular cross section are placed at strategic locations along the outside edge to deflect or redirect impinging flow away from a river bank. The vanes are oriented upstream at approximately a 30° angle with the bank. A total of five vanes were constructed out from a rock working pad on the edge of a low-flow terrace that was built out from the toe of the river bank (see Fig. 3).

- **Longitudinal stone toe protection (LSTP).** A stone toe is a continuous bank protection consisting of a stone dike placed longitudinally at or slightly streamward of the toe of an eroding bank. The cross section of the stone toe is triangular in shape. Both the rock vanes and the LSTP were placed along the outside edge of a low flood terrace adjacent the eroding bank. The LSTP was positioned to produce a smooth, curved alignment that protected the leading (upstream) part of the terrace and connected to the existing rock armor revetment next to the bridge abutment (see Fig. 4).

Fig. 3 Rock vane construction

Fig. 4 Longitudinal stone toe

Fig. 5 Live siltation construction

- **Live siltation.** Live siltation is a revegetation technique used to secure the toe of a stream bank, trap sediment, and create aquatic habitat. The basal ends of cut branches are inserted into a trench excavated parallel to the toe of the stream bank which is then backfilled with earth. The brushy tips increase bank roughness, slow velocity, and promote sediment deposition. Live siltation was incorporated in a continuous band into the leading edge of the LSTP (see Fig. 5).

Fig. 6 Overview of all measures after construction

- **Willow/Cottonwood posts and poles.** Willow and cottonwood trees may be propagated by planting cuttings. Although smaller (<4 cm (1.5 in.)) diameter cuttings (stakes) grow more vigorously than older, larger materials (posts and poles), larger materials provide better mechanical bank protection during the period of plant establishment. Dense arrays of post or poles reduce velocities near the bank or bed surface. Live willow posts were inserted into and atop the flood terrace.

An overview of all measures after completion of construction in the Fall of 2010 is shown in Fig. 6.

OBSTACLES TO IMPLEMENTATION: The main constraints were related to concerns about possible environmental and aquatic habitat impacts. Prewashing was required of any rock or gravel used and/or placed at the treatment site. Other constraints included the following:

1. The entire site was monitored for migratory salmonids. If salmonids entered the area, work was suspended, until the fish moved on.
2. Minimal disturbance to the existing banks was allowed. The permits did not allow grading the banks back nor did they allow significant filling in front of the existing embankment.
3. Monitoring of turbidity both upstream (100 ft below rock placement location) and downstream was required at all times. The turbidity was measured by taking "grab" samples. If samples showed an increase of 10 turbidity units (NTUs) above background, the work was suspended until the water was clear and the activity modified to eliminate future discharge.

PERFORMANCE EVALUATION: Evaluation consisted of (1) visual (photographic) monitoring using a webcam to record real time "live" conditions, (2) measurements of the thalweg displacement, and (3) analysis of response to winter storm events. The rock vanes and LSTP, constructed from well-graded and self-launching rock, were effective in redirecting the high flow currents away from the actively eroding bank (see Fig. 7). After a relatively short learning curve, the techniques employed were very buildable and cost-effective.

The rock vanes and LSTP functioned in a very high energy environment during the winter of 2010/2011. The willows attenuated flow velocities and helped to trap and hold sediment on the flood terrace. In spite of several instances of high water the protective structures and vegetation stayed in place and displaced erosive, high velocity flows away from the stream bank while preventing further erosion of the stream bank. Vegetation established well both atop the flood terrace and near the water's edge (refer to Fig. 7).

Fig. 7 View of project site a year later after several major floods

PROJECT COST: The Geyserville project was considered an emergency repair; it was paid for on a force account (time and materials) basis. Total installed cost of the initial work described in this case history was approximately $1 million. The materials consisted of 6,335 tons of different rock types. The cost of materials also included harvesting of all local plant materials, willows for the live siltation work, willow pole planting, and brush layers. If the job had been bid, the estimated total cost would have ranged from $500k to $650k. These lower bid estimates can be explained by a depressed regional economy and contractors who were hungry for work during this period. In addition the scope of work was fairly simple in construction technique, equipment needs, and access. The length of treated reach of stream bank was 404 ft. An alternative bank stabilization treatment such as that described in this case history tends to incur a 20–25 % markup compared to a traditional or conventional treatment, e.g., bank armor or rock riprap.

BENEFITS AND LESSONS LEARNED:

1. This project demonstrated that rock vanes can be effective in large, high energy rivers, where large RSP rock armor has traditionally been prescribed. The relatively small, self-launching stone did adjust positions as designed but essentially stayed in placed and did not move down river.

2. Construction techniques utilizing clean self-launching stone placed directly on the river bottom did not damage the bottom substrate and only incurred very small and localized spikes in turbidity. An important lesson was that the cleaner the rock, the less turbidity. Construction could then proceed without constraints by time-consuming sediment/turbidity barriers.

3. This project establishes the validity of research findings in the NCHRP Report 544 showing that redirective techniques can successfully move high velocity currents away from eroding embankments and maintain the thalweg in the relative center of the channel.

ACKNOWLEDGMENT

The design criteria for this project was derived from material developed in the guidance manual, NCHRP Report 544 – Environmentally Sensitive Channel and Bank Protection Methods, McCullah, J. and Gray, D.

The case study is adapted in part from information in the paper McCullah, Gray and Yam (2012) presented to GeoCongress 2012.

ADDITIONAL READING:

NCHRP (2005). Environmentally Sensitive Channel and Streambank Protection Measures, NCHRP Report 544. Guidance manual prepared for the Natl. Coop. Hwy. Research Program, Washington, DC.

McCullah, J., Gray, D.H., and Yam, D. (2012). Bank Stabilization at Geyserville Bridge Site Using Re-Direcive and Bioengineering Measures. *Proceedings*, GeoCongress 2012, Geotechnical Special Publication #225.

PROJECT #13: BUCKHORN MOUNTAIN

PROJECT TYPE: **Hwy cut slope repair and stabilization**

PROJECT SCALE: MEDIUM

CLIENT/OWNER: California Department of Transportation

TECHNIQUES EMPLOYED: **Modified VMSE ("soil flaps," Vegetated gabions Brushlayering Soil amendments)**

GEOGRAPHIC LOCATION: Mile post 0.1, Hwy 299 W, 0.1 mi below Buckhorn Summit Shasta Co., CA

Fig. 1 Regrading of failed slope

GEOMORPHIC SETTING: Buckhorn Mountain is an exposed decomposed granite batholith which separates Shasta and Trinity Counties in NW California. Highway 299 W traverses Buckhorn Mountain for almost 15 miles over steep mountainous terrain. The Shasta Bally Batholith granitics have a high percentage of biotite and mica-type minerals that weather easily. Weathering is extensive because the batholith is deeply fractured, thus allowing the intrusion of water.

SITE CONDITIONS AND PROBLEMS: The region can experience inches of rain in the winter punctuated by numerous freeze–thaw cycles followed by 6 months of no precipitation at all. Summer temperatures often exceed 100 °C for weeks on end. Soil texture and composition are other impediments to stabilization. The decomposed granite parent material weathers rapidly to a sandy-silt textured soil that is extremely erosive. The soils derived from the Shasta Bally Batholith are within the Hydrologic Soil Group A, based on high permeability, and have been determined to be some of the most erosive in the nation. On exposed cut slopes competent appearing bedrock quickly decomposes and ravels when exposed to moisture and freeze–thaw cycles (see Figs. 1 and 2). The soils resulting from these weathering processes are droughty. The water holding capacity, nutrient status, and organic content are also very low; therefore, these soils are difficult to revegetate.

W. Goldsmith et al., *Bioengineering Case Studies: Sustainable Stream Bank and Slope Stabilization*,
DOI 10.1007/978-1-4614-7996-3_14, © Springer Science+Business Media New York 2014

Fig. 2 Project site before treatment showing extent of erosion

Fig. 3 Seepage erosion occuring at intersection of GWT and slope face

These decomposed granite (DG) soils in upslope, undisturbed areas have extremely high infiltration rates and permeability. Therefore, the mountain has significant amounts of groundwater within. This presents another erosion mechanism that is especially problematic on steep highway cuts and fills, namely, exfiltration or "seepage erosion." This often occurs when the slope face intercepts the groundwater table (see Fig. 3).

TREATMENT OBJECTIVES AND CONSIDERATIONS: The main goal was to arrest erosion and to prevent shallow slope failures in addition to stopping sediment delivery to the highway. Previous stabilization trials on decomposed granite soils indicated that the best remedy for both of the sediment-producing mechanisms—freeze-thaw and raindrop erosion—is the same, namely, establishing and/or maintaining a good vegetative cover. Plant roots compensate for the lack of cohesion and help arrest downslope movement of soil particles while the aboveground vegetative cover dissipates raindrop impact and provides a protective cover.

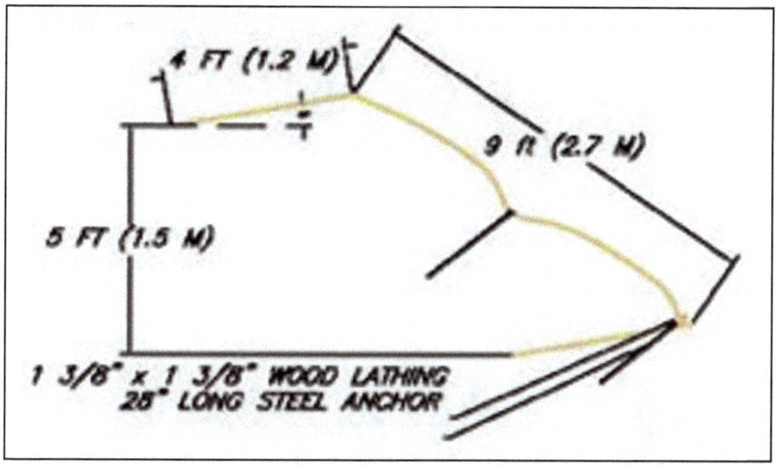

Fig. 4 Schematic illustration of external stress application on a slope

A secondary goal was to determine how to increase infiltration rates as well as water holding capacity and rooting depth. Nearby sites, that were well vegetated and stable, even though they had been disturbed 10–20 years previously, were referred to as "reference sites." The reference sites were analyzed by UC Davis researchers by measuring their infiltration rates, rooting depth, and percent OM. The plan was to amend the surface soil at the project site with a suitable compost mixture, in sufficient quantities to establish sustainably California native vegetation (early successional grasses) on these droughty, nutrient deficient soils.

Another project goal was to develop a method of utilizing a geotextile in a highly "constructable" and cost-effective manner that would stabilize the outer face of a DG soil slope. This soil lacks cohesion and is highly susceptible to sloughing in addition to seepage erosion caused by exfiltrating groundwater. The lack of cohesion in sandy granitic soils means they have low shear strengths near the surface where confining stresses are low. This condition makes granitic soils susceptible to shallow translational sliding and sloughing. The placement of an external confining stress on a slope via an anchored netting or geotextile (Ghiassian et al., 1998) can remedy this problem. A geogrid or net is placed on a slope and then pulled down tightly against the surface by means of driven line anchors. This anchoring procedure develops curvature and tension in the overlying geotextile which in turn imparts a confining stress on the ground surface. This external stress and anchoring procedure on the face of a slope is illustrated schematically in Fig. 4.

<u>TREATMENTS SELECTED</u>:

A series of biotechnical measures were adopted to help stabilize the slope against both surficial erosion and shallow translational sliding. In addition soil amendment trials were conducted with different compost mixtures to improve vegetative establishment.

Fig. 5 Gabion basket toe-wall protecting bottom of slope

- **Vegetated gabion toe-wall.** The bottom of the slope at this site was stabilized by constructing rock-filled gabion basket wall at the toe. The gabion wall was reinforced with willow poles inserted through the baskets into the soil fill at the toe (see Fig. 5).

- **Modified VMSE (soil flaps).** This treatment basically consisted of protecting the highway cut by placing a buttress fill against the cut. It was necessary, however, to construct the fill in such a way that it would be stable against shallow slope failure plus erosion and also capable of supporting a vegetative cover. The cut slope in essence was "converted" to a 1.5:1 fill slope, approximately 90 ft high, by building successive soil lifts from the highway toe to the top of the slope. **These lifts were constructed in a different manner from a conventional VMSE.** The standard construction of Veg Mech Stab Earth specifies that a geotextile (or geogrid) be laid horizontally on the terrace, the soil lift is then placed and compacted, and finally the geotextile is pulled around and over the soil lift. The process is then continued up the slope with successive lifts. Caltrans engineers believed that this conventional VMSE construction technique would be impractical for large-scale projects. Instead Caltrans District 2 engineers came up with a construction technique termed "soil flaps."

  To construct a soil flap each soil lift was approximately 5 ft high, graded at an inslope angle of 10°, and then reinforced with brush layering and coir netting. The coir netting specified was 900 g/m, which has high tensile strength and an expected durability of 5–7 years. An 8″–12″ thick compost soil admixture (25 % compost, 75 % DG soil by volume) was spread over the face of the 1.5:1 slopes prior to installation of the coir flaps. A 4 m-wide (13 ft)

Fig. 6 Construction of buttress fill protected by soil flaps and brush layers

coir geotextile roll was laid out on the bench with 2.7 m (9 ft) flapping over the face. The next lift was built and a soil flap installed; this technique was continued up the slope (see Fig. 6). For cost-effectiveness and constructability, brush layering was not installed continuously across each terrace, but instead it was installed in a staggered fashion.

To impart strength and resist downslope shear forces, a soil flap must be anchored in such a way to counter those forces. The resisting and anchoring forces were achieved by using 28 in.-long steel anchors with a 1-in. washer on the end. The anchors were inserted through predrilled, wooden lathing strips and used to secure the bottom of the flap. In addition pins were used on the coir mat every 3 ft on center in a diamond pattern. The "soil flap" approach mimics a slope stabilization procedure known as an "anchored netting system."

- **Soil amendments.** Increases in compost will increase infiltration rates, rooting depth, fertility, and water holding capacity—all critical to revegetation on these adverse slopes. Accordingly, test plots were built and monitored with varying amounts of compost admixtures (McCullah et.al. 2005). Since part of the project research involved the benefits of adding compost to the soil face and encouraging deep rooting of native grasses and willows, the design did not require compaction of the soil lifts to the standard 95+ %. However, compaction was monitored to achieve (87–93 %) with the dozer while maintaining an optimal moisture (approximately 15 %).

- **Vegetative Treatments:** The vegetative treatment included, in addition to the willow branches and compost admixture, broadcasting slow-release organic fertilizer (Biosol), mycorrhizae inoculant (AMB 100), and proven native seed mixture including Elymus glaucus, Festuca idahoensis, and

Bromus carenatus). Soil compaction can influence plant growth in a variety of ways, both good and bad. Agronomists generally recommend minimal soil compaction so as not to impede growth and development of crops and native plants. Gray (2002) discusses different strategies for optimizing soil compaction to balance engineering requirements and plant growth needs in slope protection and erosion control work.

OBSTACLES TO IMPLEMENTATION: The main obstacle to implementation was skepticism about the ability to establish vegetation on droughty, nutrient deficient, and highly erosive granitic soils. In addition there was concern about the ability to construct a buttress fill steeper than 2:1 (H:V) using largely cohesionless granitic soils that were prone to both erosion and shallow translational sliding.

PERFORMANCE EVALUATION: The slope has withstood several significant winter storms including snowfall events over the last 3 years since construction. In fact the slope was stable immediately after construction when severe storms hit the site within weeks after completion (see Fig. 7). Following a series of storms in February of 2004, which produced 5 in. of precipitation in 3 days, the slope remained stable compared to previous years. The condition and appearance of the slope 2 years after construction is shown in Fig. 8.

In 2009 Caltrans received funding for more curve correction work for safety purposes. The area was cut higher and steeper to allow for the widening. This resulted in local slope failures by seepage erosion (sapping) and slumping. Subsequent repairs consisted of placing thick layers of rock (12 to 20-ft thick) on the slope (see Fig. 9).

Fig. 7 Appearance of slope immediately after construction

Fig. 8 Condition of slope 2 years after construction

Fig. 9 Placement of rock blanket on slope

The effectiveness and constructability make anchored coir flaps an important consideration for use in non-cohesive soils subject to failures that occur as a result of saturation and mass soil movement. The total cost of the coir flap installation was less than $60,000. The construction costs included 5 workers and the use of a backhoe, excavator, bulldozer, and a water truck for 4 days, and $5,000 in materials. The field trials indicated that a 24 % compost admixture, by vol., emulated natural conditions with regard to infiltration, water holding capacity, and rooting depth.

BENEFITS AND LESSONS LEARNED:

1. This project demonstrated that vegetation establishment is greatly improved with the addition of compost. Compost incorporation may be critical to the success of projects attempting to restore vegetation and proper soil function to degraded, droughty soils.

2. No slumping or failures occurred in the initial project area, suggesting that the flap technique succeeded in neutralizing the downslope forces and destabilizing seepage.

ADDITIONAL READING:

Ghiassian, H., Gray, D.H., and Hryciw, R. (1998). *"Stabilization of Coastal Slopes by Anchored Geosynthetic Systems."* J. of Geotechnical Engr. (ASCE), Vol. 123, No. GT8, pp. 736-743.

Gray, D.H. (2002). *"Optimizing Soil Compaction and Other Strategies,"* Erosion Control **9**(6):27-34 (Sept/Oct 2002)

McCullah, J. et.al., (2005), *Effective Compost and Bioengineering Techniques for Adverse Soil Conditions,* Proceedings, International Erosion Control Association Conference, Dallas 2005.

PROJECT #14: BUCKHORN SUMMIT

PROJECT TYPE: **Hwy cut slope repair and stabilization**

PROJECT SCALE: MEDIUM

CLIENT/OWNER: California Department of Transportation

TECHNIQUES EMPLOYED:
Stepped slopes
Soil amendments
Revegetation
Brushlayers
Live staking

GEOGRAPHIC LOCATION: Hwy. 299 W, just below Buckhorn Summit, Shasta County, CA

Fig. 1 Cut-slope brushlayers

GEOMORPHIC SETTING: Buckhorn Mountain is an exposed decomposed granite batholith, which separates Shasta and Trinity Counties in NW California. Highway 299 W traverses Buckhorn Mountain for almost 15 miles over steep mountainous terrain, elevations of 1,000–3,000 ft at the summit. The Shasta Bally Batholith granitics have a high percentage of biotite and mica-type minerals that weather easily. Weathering is extensive because the batholith is deeply fractured, thus allowing the intrusion of water.

SITE CONDITIONS AND PROBLEMS: Hwy 299 is the main east/west route from Redding in the northern Central Valley to the Eureka in the North Coast, yet the terrain conditions and environmental concerns make highway construction (straightening and widening) extremely challenging. As of 2010 the CA Construction General Permit (in compliance with the Clean Water Act) became very stringent and required all new constructions to be "stabilized" before construction and permit coverage is considered "completed." Historically, steep cut slopes in these decomposed granites (DG) have been

W. Goldsmith et al., *Bioengineering Case Studies: Sustainable Stream Bank and Slope Stabilization*,
DOI 10.1007/978-1-4614-7996-3_15, © Springer Science+Business Media New York 2014

extremely difficult to vegetate; bare raveling cut slopes can also be a chronic source sediment. The region can experience 50 in. of rain in the winter followed by 6 months of no precipitation at all; summer temperatures can often exceed 100 °C for weeks on end. These conditions make the establishment of vegetation very challenging. Soil texture and composition are other impediments to stabilization. The decomposed granite parent material weathers rapidly to a sandy-silt textured soil that is extremely erosive.

On exposed cut slopes competent appearing bedrock quickly decomposes and ravels when exposed to moisture and freeze–thaw cycles. The soils resulting from these weathering processes are droughty. The water holding capacity and organic content are also very low; therefore, these soils are difficult to revegetate.

These decomposed granite (DG) soils in upslope, undisturbed areas have extremely high infiltration rates and permeability. Therefore, the mountain has significant amounts of groundwater within. This presents another erosion mechanism that is especially problematic on steep highway cuts and fills, namely, exfiltration or "seepage erosion." This often occurs when the slope face intercepts the groundwater table.

TREATMENT OBJECTIVES AND CONSIDERATIONS: The main goal was to test the viability of constructing a stepped (benched) highway cut and to conduct field trials for revegetation strategies. Stepped slopes control erosion by breaking up the slope length thereby reducing the volume and velocity of stormwater runoff (see Fig. 2).

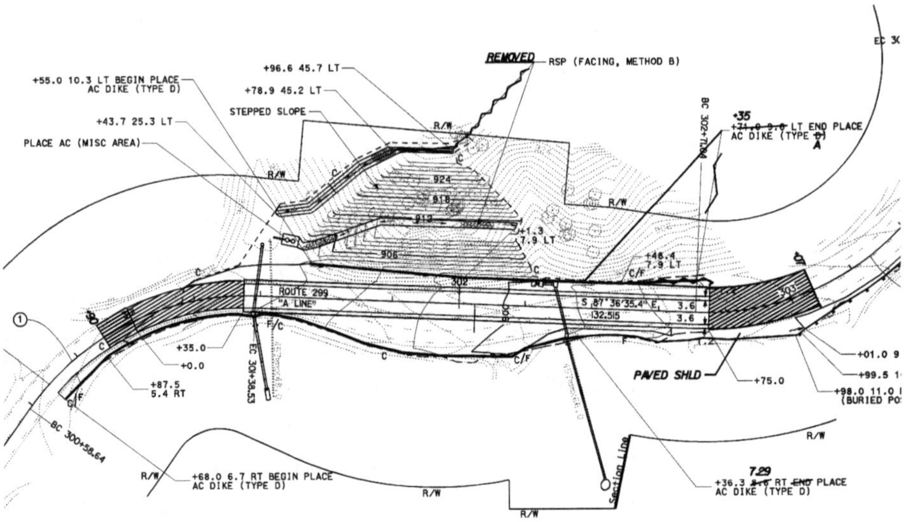

Fig. 2 As-built plans for Post Mile 1.0 showing curve correction and stepped-slope

TREATMENTS SELECTED:

- **Stepped Slope:** The slope was constructed as described in Highway Design Manual (HDM) Chapter 300, Topic 304.5 (see Fig. 3). The benches were constructed approximately 2'H by 2'V.

- **Back filling Steps:** Six of the steps, 3 above and 3 below drainage bench, in the middle of the slope were backfilled with a mixture of 25 % Certified California Compost and 75 % native DG (see Fig. 4).

- **Willow Brushlayering:** Two steps on the eastern half had 3–4″ long willow branches planted on the steps. A fiber roll (straw wattle) was placed along the outer edge of the step/bench, then the branches were placed with the basal ends pointing downward and the tips pointing up, then the step was backfilled with soil and the basal ends of the branches covered (refer to Fig. 1).

- **Landscape Treatments:** Several different treatments were tried and monitored including:

  - All the steps had native grass seeds, covered with straw mulch, planted on the horizontal surfaces.
  - Native pine trees and shrubs were planted on the steps by drilling planting holes in the bench with power augers.
  - Drip irrigation versus polymer (Dry Water). The eastern half of the slope had the plants drip irrigated by gravity from a water tank installed at the top of the slope. The western half of the slope had "Dry Water" polymer irrigation.

- **Live Willow Stakes/Slope Pins:** In late December and early January of the first winter the area experienced a snowstorm followed by a large rain event. A saturated slump occurred on the upper center of the slope (see Fig. 5). This small slide was caught on the mid-slope drainage bench, but the saturated material threatened to initiate a larger failure. Temporary repair measures were considered until the slope could be rebuilt in the summer. Plastic covers were rejected in favor of Slope Pinning with live willow stakes. Accordingly, live willow stakes 30″–36″ long × ¾″ diameter were driven into the slumping material at 3′ OC. The stakes served as "pins or dowels" that kept the slump mass in place.

Fig. 3 Post Mile 1.0 stepped slope earthwork complete on November 2002

Fig. 4 Filling bench with 25 % compost and 75 % DG admixture tried on six plots

(a) (b)

Fig. 5 Views of stepped cut. (**a**) After mid-slope slump (**b**) After pinning with live stakes. Treatment stabilized slump; no further work was required

OBSTACLES TO IMPLEMENTATION:

There were no extraordinary obstacles. There was a common false belief that mountainous, high elevations were not good locations for willow. We had to point out that *Salix* spp. often play the role of pioneering species, coming into an area naturally after fires and landslides. Then naturally occurring plants in the general vicinity were discovered and pointed out

PERFORMANCE EVALUATION

This stepped cut slope is currently well vegetated and has remained stable for 10 years. The brush layers have grown vigorously and have helped to reinforce and stabilize the slope (see Fig. 6). A small mid-slope slump that occurred in the first winter was stabilized with willow stakes (see Fig. 7). These willow stakes have persisted in stabilizing the slump without the need for further treatment. Trees grown in the compost/soil backfill have shown significantly more growth and vigor than those without benefit of this treatment. These techniques may prove to be the only feasible way to comply with the new California General Construction Permit. The site has required little to no maintenance since the first winter. Vegetation has established and grown well on the site in spite of lack of irrigation after the first year (see Fig. 8).

(a) (b)

Fig. 6 Growth of brushlayers on stepped highway cut. (**a**) after first winter (**b**) Brushlayers after 3 years

(a) (b)

Fig. 7 Long-term performance of remediation treatments 10 years later. (**a**) Brushlayers. (**b**) Willow stakes

(a) (b)

Fig. 8 Growth and establishment of vegetation on stepped-highway cut (**a**) after 3 years. (**b**) after 10 years in 2013

BENEFITS AND LESSONS LEARNED:

Stepped slopes should be considered when steep, highly erosive cut slopes are difficult to stabilize. The new Clean Water Act regulations will make it extremely difficult for Highway or County Road Department to construct slopes that are chronic sources of sedimentation. Therefore, these steep slopes with adverse soils will require extra engineering and earthwork techniques to ensure vegetation establishment.

Caltrans Division of Landscape Architecture conducted extensive research and developed Sustainable Erosion Control techniques for steep adverse soils. Stepped Slopes, RECP Flaps and RECP Wraps, and Wire Mesh Confinement (VMSE) are a few engineering-approved techniques that have Standard Special Provisions.

Backfilling the steps resulted in less erosion and much better plant establishment. Another important lesson learned was adding Certified Compost and Mycorrhizae to the steps and planting basins. It was discovered that the steps could be cost-effectively constructed, then backfilled with compost/soil/mycorrhizae admixture as the slope was being built. It was expensive to cut the benches and then go back and fill in the steps. Better to let the heavy equipment complete both tasks simultaneously.

The use of live stakes to buttress (pin) landslide material was also an important lesson.

ADDITIONAL READING:

California Department of Transportation Landscape Architecture Program, Erosion Control Toolbox, http://www.dot.ca.gov/hq/LandArch/ec/index.htm

PROJECT #15: STAFFORD

PROJECT TYPE: **Landslide repair and stabilization**

PROJECT SCALE: Medium

CLIENT/OWNER: Pacific Lumber Company

TECHNIQUES EMPLOYED: **Pole drains, Live fascines Live (mini) brush-layers**

GEOGRAPHIC LOCATION: Watershed near Stafford, Humboldt County, CA

Fig. 1 Live pole drain

GEOMORPHIC SETTING: Steep drainage basin with a watershed of about 140 acres and a length of about a mile. Sides of basin are mantled with colluvium. An intermittent stream in the drainage basin flows northward through the town of Stafford and discharges into the Eel River. The stream channel drops about 1,745 ft in elevation, in its 1-mile length. The steepest stream gradient is about 67 % (33°) near the head of the basin.

SITE CONDITIONS AND PROBLEMS: A series of nearly continuous rainstorms on the northern California coast triggered two large mudflows that originated from the head of the drainage basin and swept down the stream channel (see Fig. 2). The mudflow spread out at the base of the watershed and destroyed several homes. Most of the watershed had been clear-cut previously leaving the slopes bare and susceptible to erosion and slope failures. A timber haul road traversed the source area at the head of the mudflow. Erosion and shallow slope failures continued to occur in the watershed after the mudflow (see Fig. 3).

W. Goldsmith et al., *Bioengineering Case Studies: Sustainable Stream Bank and Slope Stabilization*,
DOI 10.1007/978-1-4614-7996-3_16, © Springer Science+Business Media New York 2014

Fig. 2 Source area of mudflows at head of drainage basin

Fig. 3 Mudflow source area with continued slope failures

TREATMENT OBJECTIVES AND CONSIDERATIONS: A debris basin was constructed at the mouth of the watershed to contain future sediment and debris resulting from continued erosion and slope failures. Another solution was to treat the head-scarp area at the upper end of the watershed, which was the main source of the mudflow, in order to minimize continued erosion. Access problems and the lack of an obvious point source favored a dispersed approach. Steep, unstable slopes in the watershed tended to negate a conventional seeding and revegetation program. Accordingly, a decision was made to try a bio-stabilization approach to treat active erosion sites in the upper watershed area using several soil bio-engineering techniques. These were sited to control ground water seepage in addition to establishing a vegetative cover that would help to prevent shallow slope failures and continued erosion in the portion of the watershed denuded by the mudflow.

TREATMENTS SELECTED:

- **Live fascines.** Fascines consisting of bundles of live willow branches were placed in shallow trenches on contour across the head or source area of the mud flow. Rather than continue the fascines all the way across the slope, in the usual manner, they were placed in selected segments that needed particular attention.
- **Live pole drains.** A pole drain resembles a large diameter fascine (refer to Figs. 1 and 4); its main function is to capture subsurface seepage and prevent the build up of pore water pressures in the soil. The pole drains were placed at strategic locations, in a herring bone fashion (see Fig. 5) up-and-down slope to maximize their effectiveness as subsurface drains.
- **Mini brushlayers.** Brushlayers were placed on small benches cut into the slope. Rather than extending all the way across the slope, they were placed on smaller benches at critical location (see Fig. 6)

Fig. 4 Live pole drain staked in place

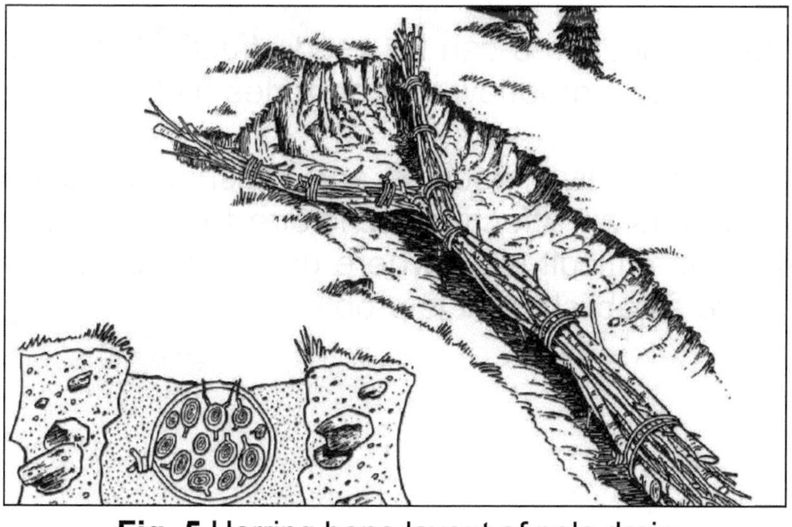

Fig. 5 Herring bone layout of pole drain

Fig. 6 Installation of mini-brushlayers across slope

OBSTACLES TO IMPLEMENTATION: The main obstacle to implementation was a reluctance to undertake any remedial action beyond the construction of a sediment basin at the mouth of the watershed. A legal suit brought by home-owners in the watershed whose homes were damaged or destroyed motivated a more proactive remediation approach to controlling erosion at its source.

PERFORMANCE EVALUATION: The remedial treatment undertaken to control erosion on the watershed slopes was limited in scope and extent. Part of the reason for this limited scope was a feeling of "Let's try it and see if it works" on the part of the project client. Results have shown that the bio-stabilization measures were effective in places where they were used. Seepage erosion was arrested in locations where the pole drains were placed, and the fascines and mini brushlayers stabilized the slope sufficiently to allow the establishment of vegetation, including native plants and shrubs (see Fig. 7).

BENEFITS AND LESSONS LEARNED:

1. Bio-stabilization measures can be used effectively to repair and stabilize areas damaged by a mudflow that denudes a steep slope.

2. The combination of live fascines and brushlayers used in conjunction with pole drains is an effective technique to minimize potential erosion and instability in difficult sites where groundwater seepage is occurring and vegetative establishment is problematic.

Fig. 7 Mudflow source area, 1 year after treatment

ADDITIONAL READING:

Huber, O.L (1997). "The Stafford, California Mudflow of December 31, 1996". Report (unpublished) prepared for the California Division of Mines and Geology, 20 pp.

Gray, D.H. (1970). Effects of forest clear-cutting on the stability of natural slopes, Bulletin of the Assoc. of Engineering Geologists, Vol VII, Nos. 1 and 2, Fall 1970.

Spitler, T. (1982). Geology and Geomorphic Features Related to Landsliding, Scotia 7.5-minute quadrangle, Humboldt County, Calif. Dept. of Conservation, Div. of Mines and Geology, OFR 82-20 SF, Scale 1:24,000

PROJECT #16: PACIFICA

PROJECT TYPE: **Landslide repair and stabilization**

PROJECT SCALE: Medium

CLIENT/OWNER: Local homeowner's association

TECHNIQUES EMPLOYED: **Slope regrading, Live fascines, Slope sub-drain, Seeding and transplanting**

GEOGRAPHIC LOCATION: Oddstad Blvd., Pacifica, San Mateo Co., CA

Fig. 1 Fascines on contour in hillside swale

GEOMORPHIC SETTING: Steep hillside with colluvium-filled bedrock hollow (see Fig. 1). These hollows are a fairly common geomorphic feature of the coast ranges in California. The hollows can become saturated during storm events thereby causing pore water pressures to rise and triggering slope failures and debris flows.

SITE CONDITIONS AND PROBLEMS: A debris flow was mobilized during an intense rainstorm in the hillside above a row of homes located at the foot of the hillside. The debris flow swept down suddenly destroying several houses and killing two children in one of the homes (see Fig. 2). Emergency treatments were undertaken in the hope of stabilizing the track of the debris flow and stopping further erosion and slope failures. The treatments consisted of removing some of the remaining colluvium in the bedrock hollow source area, constructing V-ditches across the slope, and seeding the debris flow track and adjacent slope areas with a grass mixture. These measures failed to stem continued erosion and gullying in the debris flow track. Grass establishment was minimal and largely ineffective.

W. Goldsmith et al., *Bioengineering Case Studies: Sustainable Stream Bank and Slope Stabilization*,
DOI 10.1007/978-1-4614-7996-3_17,

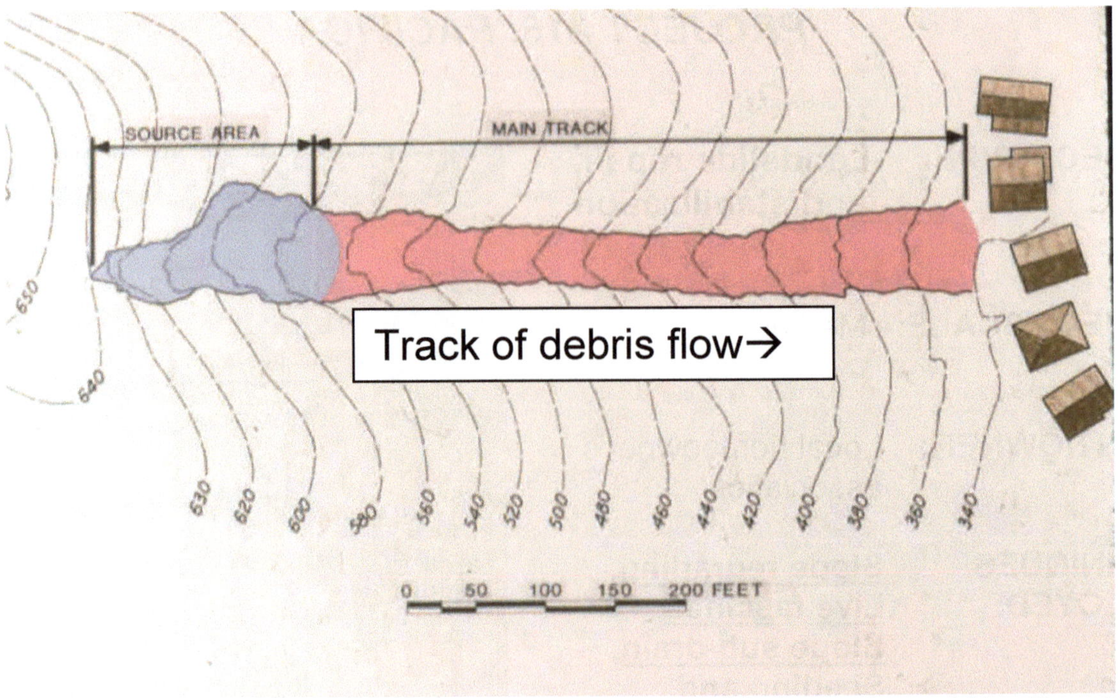

(photo courtesy of Alan Kropp)

Fig. 2 Track of debris flow that destroyed several homes

TREATMENT OBJECTIVES AND CONSIDERATIONS: The main goal was to prevent shallow slope failures and continued erosion in the debris flow track. Several alternatives were considered including complete removal of all the colluvium in the bed rock hollow versus partial removal accompanied by biotechnical stabilization of the remaining colluvium. The objective was to stabilize this material and prevent further erosion and shallow sloughing. The huge quantities of colluvium removal (necessitated by the first approach) favored the latter, less drastic solution.

TREATMENTS SELECTED:

- **Slope regrading:** The colluvium in the bed rock source area was removed to a depth of 3–5 ft above the bedrock surface. The side slopes adjacent to the landslide track were also graded back to an angle of 3:1 (H:V). The area was graded to a resemble a bowl with relatively gentle side slopes.
- **Live fascines.** Fascines consisting of bundles of live willow branches were placed in shallow trenches on contour across the bowl-shaped head or source area of the debris flow. The fascine rows were spaced apart at 6-ft vertical intervals. Rows of fascines were also continued down the track of the debris flow.
- **Slope sub-drain.** A slope sub-drain consisting of a slotted pipe wrapped in a pervious geotextile was placed in trench excavated down the center of the debris flow track. The trench was then backfilled, and the fascine rows placed across the top of the drain (see Fig. 3). The purpose of the drain was to collect and discharge subsurface ground water thus preventing the buildup of pore water pressure in the remaining colluvium.

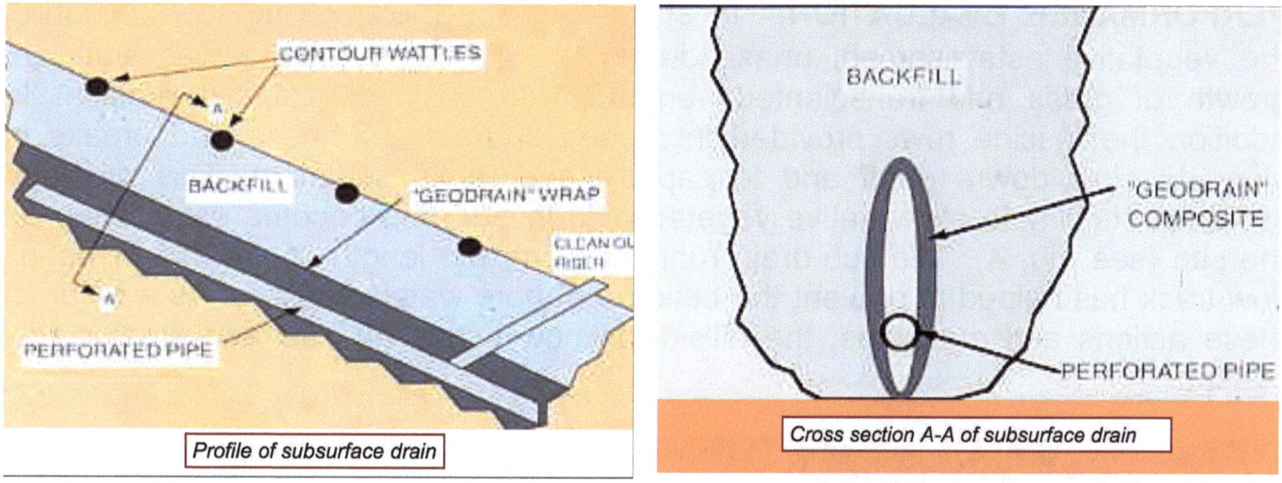

Profile of subsurface drain

Cross section A-A of subsurface drain

(drawing courtesy of Alan Kropp)

Fig. 3 Use of axial subsurface drain beneath fascines

- **Seeding and transplants.** The entire treatment area was seeded with a grass mixture of species native to the area. The same was true of transplants of woody vegetation typical of the surrounding area.

Transplant list	*Seed mixture*
Artemisia california (Sagebrush)	Berber orchardgrass (20 lbs./acre)
A. suksdorfii (Sagebrush)	Crimson Clover (3 lbs./acre)
Baccharis pilularis (Coyote Brush)	Wilton Rose Clover (10 lbs./acre)
Ceanothus cuneatus (Wild Lilac)	California Poppy (1 lb./acre)
C. intergerrimus (Wild Lilac)	Blando Brome (10 lbs./acre)
C. papillosus (Wild Lilac)	Annual Rye (5 lbs./acre)
C. parryi (Wild Lilac)	
C. ramulosis (Wild Lilac)	
C. thyrsiflorus (Wild Lilac)	
Cornus glabrata (Dogwood)	
Cupressus macrocarpa (Monterey Cypress	
Fremontodendron Calif. (Flannel Bush)	

OBSTACLES TO IMPLEMENTATION: The main obstacle to implementation was an unexpected soil moisture deficiency during the critical establishment phase. The fascines were installed and the revegetation work performed in the early fall to take advantage of the follow on winter rainfalls. Unfortunately, this phase coincided with an abnormally dry winter and below normal rainfall levels. As a result vegetation establishment and growth was not as vigorous as anticipated.

PERFORMANCE EVALUATION: In spite of a soil moisture deficiency during the vegetative establishment phase during the first winter, there was sufficient growth of grass and transplanted vegetation to minimize surficial erosion. In addition the fascine rows provided increased surface roughness and breaks in slope to slow down runoff and to capture and retain sediment. This provided sufficient stability to allow native vegetation to invade and become established on the site (see Fig. 4). The sub-drain running along the length of the former debris flow track has helped to prevent the build up of pore water pressure. As a result of these actions and outcomes, the hillside is now stable and surface erosion has ceased.

(photos courtesy of Alan Kropp)

Fig. 4 Fascine installation after installation and 1 year later

BENEFITS AND LESSONS LEARNED:

1. Bio-stabilization measures can be used effectively to repair and stabilize areas damaged by debris flows that denude steep slopes.

2. The combination of live fascines used in conjunction with a sub-drain passing under the fascine rows is an effective technique to minimize potential problems in wet sites where saturation and the build up of pore water pressures can occur.

3. Deficiencies in soil moisture caused by droughts or unexpectedly low rainfall during the critical vegetative establishment can diminish but not necessarily invalidate the effectiveness of bio-stabilization.

4. Techniques and protocols can be adopted to aid vegetative establishment under dry conditions. These may consist of temporary irrigation or the use of soil amendments, e.g., DriWater polymer granules.

ACKNOWLEDGMENT

Helpful information about this project was provided by Alan Kropp, of Alan Kropp and Associates, who directed this project. His assistance is greatly appreciated and acknowledged.

ADDITIONAL READING:

Thomas, M.R., Kropp, A. and Lucas, A. (1989). "Biotechnical stabilization of debris flow scar." Proceedings of IECA Conference XX, International Erosion Control Association.

PROJECT #17: BRANCIFORTE CREEK

PROJECT TYPE: **Riverbank repair and stabilization**

PROJECT SCALE: Medium

CLIENT/OWNER: Public Works Department Santa Cruz County CA

TECHNIQUES EMPLOYED: **Veg riprap w/ pole planting and brushlayering, VMSE, Veg gabions**

GEOGRAPHIC LOCATION: Tributary to San Lorenzo River and watershed, CA

Fig. 1 Vegetated riprap protecting bank

GEOMORPHIC SETTING: High, steep riverbank along outer bend that is subject to frequent flooding and bank scour. Branciforte Creek rises in the Santa Cruz mountains to the east and flows into the San Lorenzo River. The creek is 9-miles long and has numerous tributaries. Branciforte Creek is prime Coho salmon habitat.

SITE CONDITIONS AND PROBLEMS: High flows during large storm events resulted in erosion of a stream bank located on the downstream end of a low-radius bend with highly erodible soil. The erosion resulted in both bank scour and partial slope failure that threatened a residential structure located on the high right-descending bank (see Fig. 2).

TREATMENT OBJECTIVES AND CONSIDERATIONS: The National Marine Fisheries Service and the California Div. of Fish and Game contended that the use of riprap results in long-term habitat degradation and required the consideration of bioengineering methods for this project. Riprap and gabion designs were modified

W. Goldsmith et al., *Bioengineering Case Studies: Sustainable Stream Bank and Slope Stabilization*,
DOI 10.1007/978-1-4614-7996-3_18, © Springer Science+Business Media New York 2014

Fig. 2 Site before treatment

to include pole planting and brushlayering. The purpose of the biotechnical work was to increase slope stability, provide habitat enhancements (cover, shade, carbon, insects etc.), and improve esthetics. The willow poles and branches and cottonwood poles were harvested off-site and transported to the project site.

TREATMENTS SELECTED:

- **Vegetated riprap.** Willow and cottonwood poles were incorporated beginning approximately 1 m (3 ft) above the mean low water level, with the basal ends of the poles below or near the perennial water table (see Fig. 1). The largest rocks were placed in a toe trench and locked tightly together. Willow poles and branches were then placed in the trench with their tips protruding out over the stream. More rocks were added, and periodically, the tops of the willow poles were bent up and out towards the stream, such that approximately 60 cm (24 in.) of the terminal end was exposed to sunlight. During construction, the soil fines were placed behind and between the rocks, so that the willow cuttings would have maximum stem-to-soil contact.

- **Vegetated gabions.** Two courses of vegetated gabions were constructed at the top of the downstream end of the vegetated riprap slope. The upper gabion was constructed on top of and slightly set back from the lower gabion. Live willow poles were placed through the baskets and "toe nailed" into the slope before the baskets were filled with rock (see Fig. 3).

- **Vegetated mechanically stabilized earth (VMSE).** Vegetated mechanically stabilized earthen slope was constructed above the gabion baskets (see Fig. 4). A length of geotextile fabric was placed over each brushlayer course to provide additional reinforcement. Topsoil was then placed on top of the fabric and compacted before the fabric was pulled back over the compacted lift. Another brushlayer was placed, and this process was repeated to the top of the bank.

Fig. 3 Vegetated gabion baskets

Fig. 4 VMSE slope atop gabion wall

Fig. 5 Willow pole planting on VMSE slope at top of bank

- **Willow pole planting.** Willow poles were also inserted in rows on the VMSE slope atop the gabion wall (see Fig. 5). These provided additional root reinforcement and restraint against possible slope movement.

OBSTACLES TO IMPLEMENTATION: Some skepticism voiced initially by client about ability of biotechnical methods to protect stream bank. Habitat considerations voiced by The National Marine Fisheries Service and the California Div. of Fish and Game were needed to convince client to proceed with plan.

PERFORMANCE EVALUATION: The bank protection work was accomplished in two phases. The vegetated riprap and gabions were installed during Phase 1 of the project between January and March 2000. Additional vegetated gabions and brushlayers with soil wraps (VMSE) were placed above the Phase 1 installation in June 2000. The willow poles inserted into the riprap rooted and leafed out vigorously. The site has remained stable since construction and the soil bio-engineering installation has developed into a dense riparian buffer as native and naturalized species of woody vegetation have invaded and become well established. The site has survived several bank-full floods since construction with no signs of distress. The supple stems of willow in the riprap bend over during the flood and provide additional bank protection; the roots and stems help to lock the rock armor in place and prevent lift off (see Fig. 6).

Fig. 6 Site after treatment

LESSONS LEARNED:

1. A well-designed biotechnical system consisting of vegetated riprap, vegetated mechanically stabilized earth (VMSE), and vegetated gabions can be used to support and protect a high stream bank against both scour erosion and slope failure.

2. The bent-willow pole method of inserting live vegetation into bank armor (riprap) is superior to the joint-planting method because better (more extensive) contact is made between the live stems and underlying soil without danger of stripping the bark (as can occur when driving live stakes through rock armor).

ADDITIONAL READING:

NCHRP (2005). Environmentally Sensitive Channel and Streambank Protection Measures, NCHRP Report 544. Guidance manual prepared for the Natl. Coop. Hwy. Research Program, Washington, DC.

PROJECT #18: SAN VICENTE CREEK

PROJECT TYPE: **River bank repair and stabilization**

PROJECT SCALE: Medium

CLIENT/OWNER: Public Works Department Santa Cruz County CA

TECHNIQUES EMPLOYED: **Rock vanes, Rootwad revet, Veg. riprap-bent pole method, Live staking**

GEOGRAPHIC LOCATION: Lower San VIcente Cr. in Santa Cruz Co, nr. Davenport, CA

Fig. 1 Rootwad revetment bank protection

GEOMORPHIC SETTING: Northern California coastal stream with a 9-mile long main stem that drains a watershed with an area of 4,500 acres. The creek flows entirely within Santa Cruz County and discharges to the Pacific Ocean. Its waters rise on the west facing slopes of the Santa Cruz Mountains; its mouth is located at the community of Davenport. San Vicente creek is prime Coho salmon habitat.

SITE CONDITIONS AND PROBLEMS: Flood flows from large storms in the winter of 1998–1999 resulted in erosion of the stream bank and threatened three residential structures located on the high right-descending bank. The flood flows also undermined an existing section of rock-filled gabion baskets protecting a commercial establishment located at the lower project reach. Historically, the San Vicente Creek planform had been modified by the County Road and upstream residences. These alterations resulted in a nearly 90° turn at the upstream end of the project reach.

W. Goldsmith et al., *Bioengineering Case Studies: Sustainable Stream Bank and Slope Stabilization*,
DOI 10.1007/978-1-4614-7996-3_19,

TREATMENT OBJECTIVES AND CONSIDERATIONS: The primary purpose of the project was to stabilize the eroding stream bank along the right bank of San Vicente Creek. This project was an Emergency Watershed Protection (EWP) project funded and initially designed by USDA and NRCS and administered by the County of Santa Cruz. This project was completed in two phases. The project required environmental permits and approvals from California Department of Fish and Game and National Marine Fisheries (NMFS) because of the endangered species habitat (Coho salmon, steelhead, and red-legged frogs). NMFS and Fish and Game contended that the use of gabions and riprap results in long-term habitat degradation and required the County of Santa Cruz to consider bioengineering methods and include habitat enhancement and conservation measures for this project. The purpose of the biotechnical work was to increase slope stability, provide habitat enhancements (cover, shade, insects, carbon etc.), and improve esthetics.

TREATMENTS SELECTED:

- **Vegetated riprap (bent-pole method).** The riprap design was modified to include willow pole planting and brushlayering. The riprap apron, designed by NRCS, was 4 m (12 ft) high and 1 m (3 ft) thick, and placed on a 1V:2H smooth graded slope. A 1–1.5 m (3–5 ft) deep toe trench was excavated and the entire structure was underlain with geotextile fabric. The modifications to the traditional riprap installation involved placing the live willow branches and willow poles in contact with the soil embankment. Maximizing soil contact along the branch was desired to increase establishment (see Figs. 1 and 2). The fabric was placed on the 1V:2H slopes in preparation for the rock. Then slices were cut into the fabric beginning approximately 1 m (3 ft) above the mean low water level, so willow poles could be inserted through the fabric. The willow poles were laid flat on the smooth graded embankment with the top part of the pole on top of the fabric and the bottom part below. The basal ends of the poles were located below or near the perennial water table. The largest foundation rocks were placed in the toe trench and locked tightly together. Some willow poles and branches were placed in the trench with their tip protruding out over the stream. More rock was added to the structure and periodically the tops of the willow poles were bent up and out such that approximately 60 cm (24 in.) of the terminal end was exposed to sunlight (see Fig. 3).

Fig. 2 Vegetated rock revetment (bent willow pole method)

Fig. 3 Willow branches protruding through rock armor

- **Rootwads.** Rootwads were incorporated into the rock revetment to provide additional resistance to flows on the outside bend of the stream. Rootwads consisting of the basal trunk of a tree and associated root ball were laid in trenches excavated into the bank and then covered with soil and rock (see Fig. 4). Rootwads are oriented upstream into the stream flow where they also increase roughness and tend to cause sediment deposition (see Fig. 5).

Fig. 4 Installation of rootwads

Fig. 5 Site after placement of rootwads and rock revetment

- **Rock vane.** A severe bend in the bankline directed flow against the bank. This impinging flow was causing serious scour and erosion. A rock vane was constructed 12 m (40 ft) upstream of the critically eroded bank to deflect the stream flow away from the bank and alleviate erosive hydraulic forces. Graded, poorly sorted (self-launching) stone was specified for the vane since excavation was not allowed as required by the permit. The permit allowed placement of boulders in the stream and on the existing substrate only.

The use of self-launching stone and the upstream orientation of the vane facilitated the use of much smaller rock than the stone specified for the riprap. The vane was approximately 4.5 m (15 ft) long and was angled 30° upstream. The keyway was excavated into the bank 1.8 m (6 ft) and vegetated with pole plantings as it was backfilled with stone. The crest of the vane was sloped at a grade of 1V:5H from the average high water elevation at the bank down to the channel bottom.

OBSTACLES TO IMPLEMENTATION: Some skepticism voiced initially by client about ability biotechnical methods to protect stream bank. Habitat considerations voiced by The National Marine Fisheries Service and the California Div. of Fish and Game were needed to convince client to proceed with plan.

PERFORMANCE EVALUATION: The bank protection work was accomplished in two phases. Vegetated riprap protection was installed along the project reach during Phase 1 in 1998–1999. Rock Vane and Rootwad Revetment structures were designed and installed during Phase 2 in 2001. The willow poles inserted into the riprap (using the bent willow pole method) rooted and leafed out vigorously (refer to Figs. 2 and 3). The site has remained stable since construction and the soil bioengineering installation has developed into a dense riparian buffer as native and naturalized species of woody vegetation have invaded and become well established. The site has survived several bankfull floods since construction with no signs of distress. The rock vane and rootwads have successfully deflected and redirected erosive flows away from the bank.

LESSONS LEARNED:

1. A well-designed biotechnical system consisting of vegetated riprap, rock vanes, and rootwads can be used to protect a stream bank against scour and deflect impinging flows away from the bank.

2. The bent-willow pole method of inserting live vegetation into bank armor (riprap) is superior to the joint-planting method because better (more extensive) contact is made between the live stems and underlying soil without danger of stripping the bark (as can occur when driving live stakes through rock armor).

ADDITIONAL READING:

NCHRP (2005). Environmentally Sensitive Channel and Streambank Protection Measures, NCHRP Report 544. Guidance manual prepared for the Natl. Coop. Hwy. Research Program, Washington, DC.

PROJECT #19: OPAL CLIFFS

PROJECT TYPE: **Bluff repair and stabilization**

PROJECT SCALE: Small

CLIENT/OWNER: Local homeowner

TECHNIQUES EMPLOYED: **Anchored TRM, Live staking, Hydroseeding**

GEOGRAPHIC LOCATION: Opal Cliffs Drive, San Cruz, CA

Fig. 1 Anchored TRM on steep cliff

GEOMORPHIC SETTING: Sea cliff subject to active erosion and periodic retreat. The Opal Cliffs-Capitola reach is characterized by an irregular shoreline backed by cliffs ranging from 35 to 75 ft in height. The coastal cliffs throughout most of the city of Santa Cruz and neighboring Capitola are composed of erodible sediments of the Purisma Formation (siltstone and sandstone) along with the Santa Cruz Mudstone. These sedimentary rocks are often capped by 6–20 ft of unconsolidated marine and non-marine terrace deposits. The horizontal bedrock stratigraphy is easily visible in exposed or bare sections of the cliffs.

SITE CONDITIONS AND PROBLEMS: The Opal Cliffs section of the coastline is at high risk from erosion; the narrow beaches provide little protection from wave attack at the base of the cliff. Urbanization and house building at the cliff top causes further subaerial erosion threats, viz., concentrated runoff and subsurface seepage from streets, drains, downspouts, and excessive lawn watering. The face of the cliff at the location of the residence was actively eroding and retreating (see Fig. 2).

W. Goldsmith et al., *Bioengineering Case Studies: Sustainable Stream Bank and Slope Stabilization*,
DOI 10.1007/978-1-4614-7996-3_20, © Springer Science+Business Media New York 2014

Fig. 2 Coastal bluff before treatment

TREATMENT OBJECTIVES AND CONSIDERATIONS: As a result of the high erosion risk from wave attack, large sections of the sea cliff are protected by structures including sea walls and rock revetments. The base of the cliff at this location is protected by a sea wall and rock armor. The main problem appeared to be erosion and shallow sloughing at the face of the bluff. The goal was to arrest this erosion and stabilize the face. Adjacent stable sections of the cliff were well vegetated. If the face of the bluff was stabilized sufficiently, this would allow the establishment of vegetation which in turn would help control erosion problems. There was insufficient evidence of emergent seepage at the bluff face to warrant installation of horizontal drains (hydraugers) on the bluff face or a trench interceptor drain atop the bluff.

TREATMENTS SELECTED:

- **Anchored TRM**: A three-dimensional, turf reinforcement mat (TRM) was draped over the cliff face and anchored (nailed) to the slope by driving pins with large washers through the TRM into the slope (refer to Figs. 1 and 3). Workers rappelled down the bluff face on ropes to drive the pins. Two-foot long anchor pins were inserted on 3-ft centers (see Fig. 4).

Fig. 3 Anchoring TRM to slope face

Fig. 4 TRM attached to slope face with anchor pins

- **Live stakes.** Once the TRM was securely fastened to the slope, live willow stakes were inserted through the mat into the slope (see Fig. 5). The near vertical inclination of the cliff face required that this installation be done by workers suspended by a rope and sling system from the top of the bluff.

- **Hydroseeding.** A specially designed mixture was sprayed on to and through the turf reinforcement mat by workers suspended on ropes from the top using handheld spray nozzles (see Fig. 6). The mixture consisted of *Flexterra* @ 4000#/ac, native grass seeds (Maritime mix), *Jumpstart* (Profile) humic acid, and *BioPrime* (Profile) mychorrizae.

Fig. 5 Inserting live stakes through TRM

Fig. 6 Spraying hydroseed mix on/thru TRM

Fig. 7 Views of bluff site immediately after treatment

OBSTACLES TO IMPLEMENTATION: The main obstacle to implementation was the need to work on a near-vertical slope which required the use of ropes and slings. A system was quickly developed, however, that overcame this challenge without the need for an elaborate scaffolding or support system.

PERFORMANCE EVALUATION: A vegetative cover was soon established that was initially held in place by the anchored TRM. This provided enough time for the live stakes to root and gain a toehold on the steep slope. Erosion and shallow sloughing on the face has been arrested (see Fig. 7). Only time will tell whether this treatment will suffice in the long run, or whether other measures such horizontal drains and/or an interceptor trench drain will be required.

BENEFITS AND LESSONS LEARNED:

1. Bio-stabilization measures can be used effectively to repair and stabilize the face of coastal bluffs. Special anchoring provisions may be required when attaching a TRM to a near vertical slope face.

2. The combination of an anchored TRM and live staking were used in conjunction with a site adapted hydroseeding mix to stabilize a bluff face.

ADDITIONAL READING:

Brabb, E. (1997). Geologic Map of Santa Cruz County. US Geological Survey, Menlo Park, CA

PROJECT #20: LOWER SULFUR CREEK

PROJECT TYPE: Salmonid Stream Restoration

PROJECT SCALE: Moderate, approximately 1 mile of stream realignment and habitat restoration

CLIENT/OWNER: Sacramento Watershed Action Group (SWAG) and Co-ord. Resource Management Group

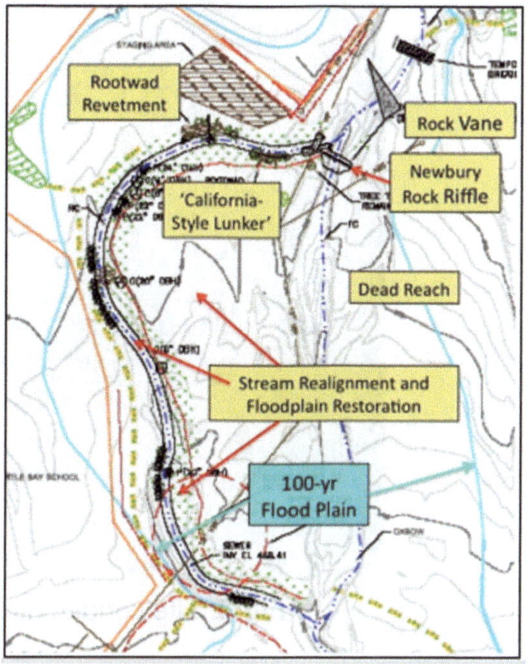

Fig. 1 – Map of site features and treatments

TECHNIQUES EMPLOYED:

- **Willow and Cottonwood Pole Planting,**
- **LWD and Rock Habitat Structures,**
- **LSTP w/ Live Siltation,**
- **Newbury Rock Riffle,**
- **Rock Vanes,**
- **Stream Realignment**

GEOGRAPHIC LOCATION: Turtle Bay/Redding Arboretum, North Redding, CA.

GEOMORPHIC SETTING: Sulfur Creek is a small, seasonal tributary stream to the Sacramento River. The watershed comprises approximately 3,000-ac with 7-mile stream length. The Sulfur Creek Watershed Analysis and Action Plan (SCWAAP), 1996, documented that the lower 2 miles of Creek can provide valuable salmonid spawning and rearing habitat if stream form and function can be restored (Fig. 1).

SITE CONDITIONS AND PROBLEMS: The entire watershed, especially the in-stream sections, has been severely impacted by hydraulic mining and dredger mining in circa 1800s and gravel mining and road/highway

W. Goldsmith et al., *Bioengineering Case Studies: Sustainable Stream Bank and Slope Stabilization*,
DOI 10.1007/978-1-4614-7996-3_21, © Springer Science+Business Media New York 2014

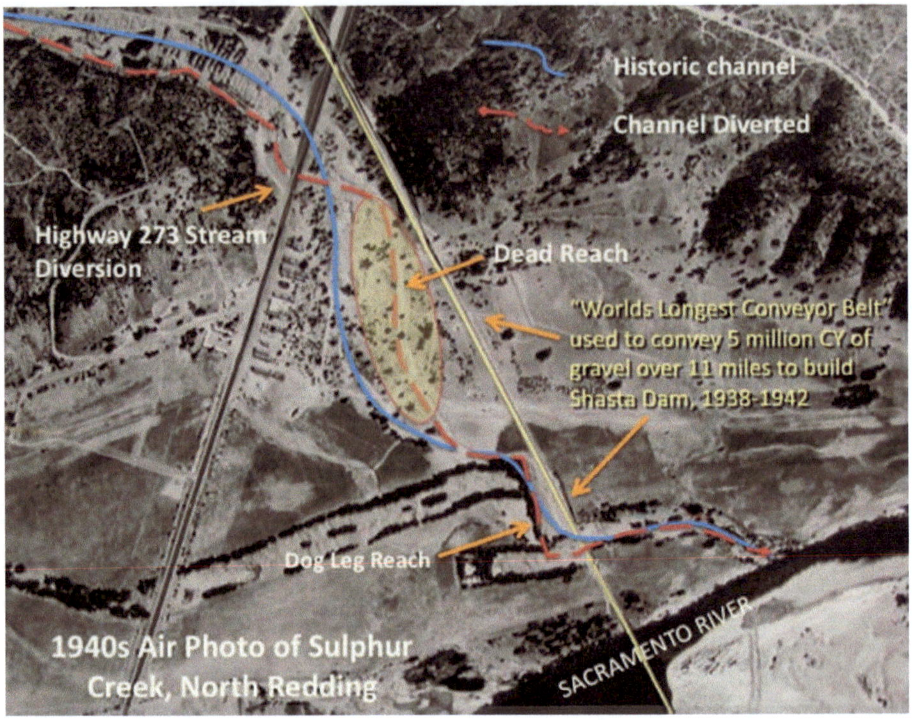

Fig. 2 Site conditions in the Lower Sulfur Creek Watershed

building in the early-mid 1900s (see Fig. 2) Lower Sulfur Creek (approximately 1 mile) runs through the Redding Arboretum. This section had been dredged for gold (circa 1850s and 1920s), and some of the dredger piles had been subsequently mined for gravel (circa 1930–1950) thereby removing the gravel-sized substrate from the stream system and leaving cobble- and small boulder-sized piles of rock in the stream and flood plain. A very significant land use impact occurred in 1934 when Highway 273 was build. The new box culvert diverted the stream out of its' historic channel and into the adjoining oak/savannah woodland.

TREATMENT OBJECTIVES AND CONSIDERATIONS:

Prior to restoration the stream in this reach is severely aggraded. The stream gradient in this lower reach is very low—an ancient alluvial fan—therefore the streams ability to transport sediment and bedload is low. The historic land uses have exacerbated this problem because the stream has been diverted out of its original channel by the upstream highway box culvert, and it has been diverted around and through the gravel mining tailings. Early gold mining "turned the channel upside-down"; then gravel mining in the mid 1900s removed the gravel-sized alluvium. Subsequently, the creek was diverted into modified dredger areas. In summary, the pre-restoration condition was characterized by a channel with a convex-up stream bottom, choked with large cobble, severe bank erosion, and little to no bank vegetation. One 1,000-ft section of the stream was nicknamed "the dead reach" for all the young escaping salmonid fry that died as the stream dried up (went subsurface) episodically (see Fig. 3).

Fig. 3 The "Dead Reach" in Lower Sulfur Creek

The primary objective for improving stream function was to realign and mimic it's historic plan form while excavating the excess bedload material. The secondary goal was to increase instream and riparian habitat. The plan for fisheries improvement was threefold:

1. Remove obstacles to anadramous fish migration and improve spawning habitat
2. Provide rearing habitat
3. Improve late Spring flows to allow "fish escapement"

Simultaneously, the SWAAP designated and prioritized watershed-wide projects intended to reduce erosion and sedimentation. The entire Sulfur Creek Restoration effort implemented by Sacramento Watersheds Action Group (SWAG) was conducted over an 11-year period, 1996–2007.

TREATMENTS SELECTED:

- **Stream Realignment and Floodplain Restoration:** In 2002 SWAG and the Sulfur Creek CCRMP received a $160,000 grant from Cantara Council to design, permit, and realign approximately 2,000 ft of stream. A new low flow channel was excavated to divert Sulfur Creek back into it's historic channel. The historic stream dimensions, e.g., Bankfull width, slope, and sinuosity, as determined from a "reference reach," were used to guide design and implementation. The inner bend(s) of the realigned reach were excavated to allow floodwaters to access the floodplain.
- **Newbury Rock Riffle:** In order to reduce the risk of flooding in the realigned reach, and ensure the stream had a sustained low flow for fish escapement, a split flow was desired at the upstream diversion. Flood stage was determined at about 3,000 cfs and at approximately 500 cfs the stream was designed to flow into both the degraded reach and the new realigned reach. The split flow was achieved with a rock weir designed as a Newbury Rock Riffle. The riffle crest was built 1-ft higher than the bottom of the low flow realigned channel (see Fig. 4).

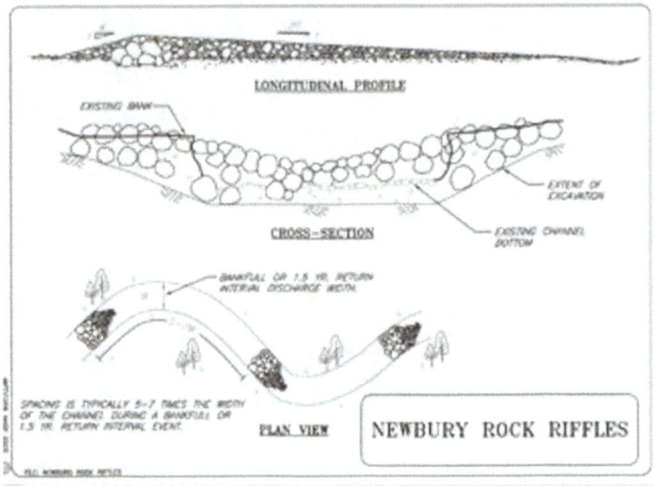

Fig. 4 Schematic diagram of Newbury Rock Riffle

Fig. 5 LWD structure constructed from logs/rootwads, boulders and pole plantings for vegetative anchoring

- **Large Woody Debris (LWD) and Rock Structures:** The realigned reach had LWD structures constructed at strategic places along the outer bends—primarily locations anticipated to receive impinging flows—thus fostering the development of scour/resting pools. The structures were constructed of large wood (rootwads), large rock, and anchored with deeply planted willow and cottonwood poles (see Fig. 5).
- **Rock Vane:** A rock vane was built approximately 300 ft upstream of the realignment project. The vane was intended to protect a mature oak tree and begin to "re-direct" the stream flow into the realigned reach. This vane, built in 2003, was the first rock vane designed and built by the author. It has been thoroughly monitored and photographed over the last 10 years. The oak tree subsequently burned down during a wildfire, however the rock vane continues re-direct flows away from the exposed bank while promoting instream habitat (see Fig. 6).

- **Longitudinal Peaked Stone Toe with Live Siltation:** Longitudinal Peaked Stone Toe (LPSTP) are generally built low. The willow branches of Live Siltationare arrayed in such a way to provide local roughness. LPSTP with Live Siltation can not only arrest outer bank erosion, but often lead to deposition in those areas. By September 23, 2003, the restoration effort had run out of grant money and time (the NPDES/CaDFG Permits allowed work until October only). Since there were insufficient funds or time to "layback" the tailings along the outer bend, an alternative construction technique was employed. The outer bend material was excavated nearly vertical. Then a rock toe about 3-ft high (designed to a height near average annual or bankfull elevation) was built longitudinally along the bend (see Fig. 7). Before the vertical cut could collapse, willow branches (Live Siltation) were quickly and carefully placed behind the rock. The branches were 6–9 ft long to insure they were placed as deep as possible for establishment. Earthen debris and tailings from the vertical cut were then raked down to backfill behind the LPSTP and around the willow (see Fig. 8).

Fig. 6 Rock vane built at beginning of project reach—to protect Oak tree left descending bank and to direct flows toward Newbury riffle and restored reach

Fig. 7 Dogleg excavation looking downstream. Outer bend is nearly vertical; rock and log toe are being placed in preparation for "Live Siltation"

Fig. 8 Dogleg rock toe and "live siltation" 7 years after construction

OBSTACLES TO IMPLEMENTATION:

The biggest obstacle to this project was the permitting process. The CEQA rules and guidelines were applied to this project as if it was a development instead of a self-mitigating restoration. All parties, the City of Redding and SWAG were inexperienced, therefore it took about 2 years and almost ½ of the restoration/implementation budget to permit the project. Eventually SWAG had to complete City/FEMA Flood Map revisions even though the restoration involved removing tons of gravel tailings from the floodplain.

Urban stream restoration activities seem to make regional flood planners nervous—unfortunately many commonly-used tools such as HEC-RAS do not accurately reflect the impacts of localized activities such as willow planting, rootwad/large woody material placement, or redirective methods such as rock vanes. HEC-RAS assumes that the channel bottom is essentially "fixed" while the restorationist wants scour pools, riffles etc. The localized scour and low flow incisions quite often compensate for placing some material in the floodplain.

Secondarily, State and Federal laws, which are interpreted and enforced by inexperienced regulators, is extremely problematic. For example, at one time a State Water Quality regulator advised the project supervisor to put silt fences in the stream and to line the channel bottom with erosion control mats. The regulator apparently failed to understand the nature of restoring aquatic and salmonid habitat. Future stream restoration projects which were regulated by State and Federal Agency staff familiar with, and supportive of, restoration activities went much smoother and resulted in maximum ecosystem benefits.

PERFORMANCE EVALUATION

The Lower Sulfur Creek project is relatively complex and required careful coordination among its different elements and components. The system performed well during and after intense rainstorms and high velocity stream flows following construction. In spite of several instances of high water the protective structures and vegetation stayed in place and displaced erosive, high velocity flows away from the stream bank while preventing further erosion of the stream bank. Low flows are now diverted to the restored channel by a Newbury Rock Riffle on left diverts low flows to restored channel (see Fig. 9). Vegetation established well both atop the flood terrace and near the water's edge. Channel degradation appears to have been arrested and sediment is being flushed through the creek to the Sacramento River. The salmonid fish habitat has been restored and frye are no longer trapped during low flows which are now diverted to the restored channel.

BENEFITS AND LESSONS LEARNED

The restoration of lower Sulfur Creek required the use of heavy equipment for restoration. The idea that heavy equipment could be used in or near seasonally dry or flowing streams, by skilled operators and careful planning was a paradigm shift that was achieved by tactful communication, frequent demonstrations, and performance monitoring. The fact that the Salmonid species returned was definitely beneficial.

One very important method (BMP) was developed and demonstrated for working in seasonally dry gravel-bedded streams, namely "Washing of Fines." Running heavy equipment in a stream channel during the dry season will ultimately bring the finer sand and silt particles to the surface. These "fines" can cause water quality exceedance when the first winter storms come.

Fig. 9 Newbury Riffle *on left* diverts low flows to restored channel *on right* 6 years after project completion

Washing fines is an "in-channel" sediment control method which uses water, either from a water truck or hydrant, to wash any stream fines, brought to the surface of the channel bed during restoration, back into the deep interstitial spaces of the gravel and cobbles, leaving clean cobble and gravel on the bed surface (see Fig. 10).

Fig. 10 Washing fines using high pressure hose from a water truck

SWAG monitored the stream turbidity during the first precipitation events resulting in stream flow. The results were astonishing because the practice of "washing fines" most often resulted in reduced turbidity when comparing upstream and downstream values. Sulfur Creek, being an urban stream with extensive upstream disturbances, could actually be "cleaned up" by Washing Fines in the restored construction reaches.

ADDITIONAL READING:

NCHRP (2005). Environmentally Sensitive Channel and Streambank Protection Measures, NCHRP Report 544. Guidance manual prepared for the Natl. Coop. Hwy. Research Program, Washington, DC

E-SenSS - Environmentally Sensitive Streambank Stabilization Techniques, 2005, Salix Applied Earthcare, Redding, CA

PROJECT #21: SECRET CANYON

PROJECT TYPE: Road fill stabilization
Stream bank restoration
Riparian zone reveg.

PROJECT SCALE: Moderate, 1-ac, 200′ of stream, remove and stabilize 10,000 CY of road spoil

CLIENT/OWNER: Sulfur Creek Coord. Resource Mgt. Group (CRMP), Sacramento Watershed Action Grp. (SWAG)

TECHNIQUES EMPLOYED:
- **Rock and fiber-roll toe with brushlayering**
- **Pole planting,**
- **Woven fences**

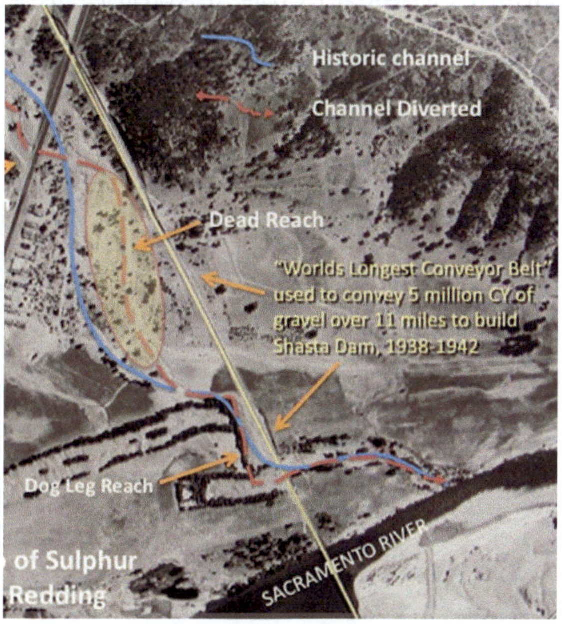

Fig. 1 Plan view of Secret Cyn site

GEOGRAPHIC LOCATION: Redding Arboretum, Redding, CA

GEOMORPHIC SETTING: Secret Canyon drains a small intermittent stream tributary to Sulfur Creek, which is a small, seasonal tributary stream to the Sacramento River (see Fig. 1). The canyon is intersected by "the Beltline," historically touted as the world's longest conveyor belt, built in 1938, to carry rock over 11 miles to build Shasta Dam (see Fig. 2).

SITE CONDITIONS AND PROBLEMS: Secret Canyon was filled in as part of the Beltline Road construction. It was a cut and fill operation. As the Beltline Road was cut through a steep hillslope, the spoil was used to fill in the adjacent tributary stream and canyon. At the time of construction (circa 1937) a 36″-CMP culvert was placed in the bottom of the canyon and then filled with an estimated 6000–10,000 CY of spoil.

W. Goldsmith et al., *Bioengineering Case Studies: Sustainable Stream Bank and Slope Stabilization*,
DOI 10.1007/978-1-4614-7996-3_22, © Springer Science+Business Media New York 2014

The 1998 Sulfur Creek Watershed Analysis and Action Plan (SCWAAP) documented that the 60-year-old culvert was failing and beltline road crossing had a high potential to erode and dump 4,000 CY of sediment into the then partially restored Sulfur Creek and the spawning areas in the Sacramento River (see Fig. 3).

Fig. 2 The Coveyor Belt over No. Market Street, circa 1938

Fig. 3 View of failing culvert before "daylighting" began

TREATMENT OBJECTIVES AND CONSIDERATIONS:

The main goal was to Daylight the stream, i.e., remove the anthropogenic road fill and replace it on the adjacent cut—slope where it came from. The total amount of fill removed from the crossing was about 10,000 cu.yds. The limited size of the spoil area required the fill be placed at 1.5:1 or steeper, so special BMPs and bioengineering methods would be employed to ensure the fill was stable.

TREATMENTS SELECTED:

- **Landform Restoration:** Landform restoration principles, e.g., restoring or mimicking natural slope and channel drainage patterns were utilized (Schor and Gray 2007)
- **Live Fascines and Willow Fences:** Willow fascines and fiber rolls are also known as willow wattles and straw wattles. The term willow wattles referring to a bundle of willow branches was first coined by Charles Kraebel, a Landscape Architect working for the US Forest Service in the late 1930s. These bundles are normally staked on contour across slopes to reduce apparent slope length and help prevent erosion and shallow sloughing. Live Smiles, are woven willow fences, usually less than 10 ft long, formed into a "smile" (catenary curve) and used to hold slumping soils on the slopes. In addition to structural stability, these measures provide, they also have the advantage of growing, sending down roots which reinforce the soil and stabilize a slope. Live Smiles have been employed successfully and are documented by David Polster, a Bioengineer from British Columbia (see Fig. 4).

Fig. 4 Live woven-willow fences stabilizing slumping slope in glacial till along a road in British Columbia

- **Live Brushlayering combined with Cottonwood and willow Poles:** The areas adjacent to the newly daylighted stream were stabilized with brushlayering and pole planting. The branches and poles were harvested from the nearby areas and then soaked for 10 days prior to installing them as deeply as possible.
- **Native Grass seeding and shrub planting** (containers): All areas disturbed by construction were revegetated with grass seeds native to the area (Elymus glaucus, Nassella pulchra), native shubs, and mulched with straw mulch.
- **Mycorrhizal fungi:** Mycorrhizae was added to a 1,000 sq. ft plot (20′ × 50′) on the fill just before seeding and mulching with native grasses. The practitioners were quite certain that there was no naturally occurring mycorrhizae in the 60-year-old road fill and wanted to see if adding Arbuscular Mycorrhizae AM-120 at 40#s per acre would enhance native grass establishment.

OBSTACLES TO IMPLEMENTATION:

SWAG and the Shasta College Natural Resources Department cooperated on this project. SWAG received a $9,000 grant from CA Department of Water Resources, Urban Streams Restoration Program (USRP), and SWAG would reimburse the College for fuel and expenses. However, the grant request did not pay for the amount of road fill required for removal. The estimated heavy equipment time was underestimated by about 50 %. Additionally, as the excavation work proceeded increasingly large boulder-sized rocks were encountered in the road fill. The College's D-6 and D-4 sized bulldozers were undersized to accomplish the work effectively. SWAG had to rent a large D-8 dozer and a 50,000 kg excavator, and several dedicated Shasta College students worked many extra hours to complete the earthwork.

PERFORMANCE EVALUATION:

The Live Smiles proved extremely effective in holding back large amounts of saturated shallow-slumping material. The key to success was that the willow stakes and woven branches were shaped like a catenary curve, a parabolic shape that maximized strength (refer to Fig. 5). Photos of the "Live Smiles" after construction and 9 years later are shown in Fig. 6.

The daylighted stream has been functioning for over 10 years now, and no erosion has ever been observed. The biotechnical plantings in the stream corridor were very successful, quite probably over 90 % success. While the plant establishment rate was not measured, the entire stream bank was completely covered in willow and cottonwood by the second year.

The bioengineering practices, willow fascines, live smiles, and live staking were critically important in stabilizing the slumps and arresting gullies before they became more serious and persistent. These practices "held the slope together" until deeply rooting native grasses became established in the second year (see Fig. 7).

The mycorrhizal plot did not have noticeably more vigorous native grasses, but almost no noxious nonnative weeds were observed compared to the untreated areas. Over the ensuing several years, SWAG has incorporated this treatment into every restoration/revegetation project (greater than15 acres). The benefits have been significant with regard to vigorous native grass stands with little to no noxious weed infestations.

Fig. 5 Installation of "Live Smile" to retain a small slump that occurred as a result of runoff saturating a steep fill slope

(a) (b)

Fig. 6 Live smiles stabilizing a slope. (**a**) During first winter, (**b**) After 9 years, the willows are also providing habitat

(a) (b)

Fig. 7 Views of Secret Canyon after bioengineering restoration. (**a**) Daylighted stream during first spring. (**b**) Canyon after 9 years with fill slope in background

BENEFITS AND LESSONS LEARNED:

The community and Turtle Bay Arboretum volunteers rallied to complete the work. The Arboretum Nursery propagated native plants and volunteers then planted the plants during "community planting days." These activities helped spread awareness of value of bioengineering restoration methods.

ADDITIONAL READING:

Kraebel, C. (1936). *Ersion contol on mountain roads,* USDA Circular #380. 43 pp.

Polster, D. (2008). Bioengineering for Land Restoration and Slope Stabilization, unpublished short course notes, Duncan, British Columbia, CAN.

Schor, H.J. and Gray, D.H. (2007). Landforming: An Environmental Approach to Hillside Development, Watershed Restoration, and Mine Reclamation. John Wiley and Sons, Hoboken, NJ.

Eastern USA

PROJECT #22: GREENFIELD ROAD

PROJECT TYPE: **Hwy cut slope repair and stabilization**

PROJECT SCALE: Large

CLIENT/OWNER: Massachusetts Department of Transportation

TECHNIQUES EMPLOYED: **Live brushlayers, Live fascines, Coir netting**

GEOGRAPHIC LOCATION: Greenfield Road near Hwy. 2 and Colrain, MA

Fig. 1 Brushlayer buttress fill

GEOMORPHIC SETTING: Residual soil with variable thickness overlying a quartz–mica schist bedrock exposed in a road cut along a scenic state highway in north west Massachusetts.

SITE CONDITIONS AND PROBLEMS: Roadway widening and encroachment during upgrading of two-lane highway resulted in high, over-steepened road cuts with marginal stability. Shallow surface failures (sloughing of soil along underlying bedrock contact) in addition to deep seated slumps (where little or no bedrock exposure) are shown in Fig. 2. Cut slope was constructed at 1.5(H) to1.0(V) grade, approximately 1,200 ft long with 20- to 60-ft high slopes. Unfavorable bedrock attitude with bedding planes and soil/bedrock contact dipping into cut. Water table present in overlying soil and active groundwater seepage from slope face and fractures in underlying bedrock. Stabilization required interception of ground water emerging from cut to prevent seepage erosion.

W. Goldsmith et al., *Bioengineering Case Studies: Sustainable Stream Bank and Slope Stabilization*,
DOI 10.1007/978-1-4614-7996-3_23, © Springer Science+Business Media New York 2014

Fig. 2 Cut slope failures along highway

TREATMENT OBJECTIVES AND CONSIDERATIONS: The fractured nature of the bedrock, its uncertain depth, and unfavorable attitude ruled out the use of structural retaining walls resting on the bedrock interface. A drained rock buttress at the toe of the cut would have satisfied mass stability requirements but would have left the top part of the cut exposed and vulnerable to seepage erosion. This limitation could have been overcome by running the rock all the way up the slope in the form of a rock blanket. This solution, however, was visually stark and environmentally incompatible.

Visual and esthetic concerns—strongly expressed by local residents—favored treatments that were visually nonintrusive and that blended in with natural surroundings. A drained, earthen brushlayer fill or facing was proposed initially to satisfy this requirement. Project engineers were concerned about the adequacy of shear resistance and drainage at the base of an earthen fill. A compromise solution was eventually devised that consisted of placing a 10-ft high rock buttress at the toe of the cut which in turn supported a drained, brushlayer fill above. Shear strength tests on the fill material in combination with slope stability analyses were employed to design this hybrid system. The stability analyses showed that the rock buttress at the bottom intercepted the critical failure surface which passed through the toe of the slope.

TREATMENTS SELECTED:

- Drained, earthen **brushlayer fill** (or facing) supported on 10-high, crushed rock buttress at toe of cut slope was employed in portions of the cut where no bedrock was exposed (refer to Figs. 1 and 3). A drainage course was placed behind brushlayer fill to intercept and divert seepage down into rock toe at base.

- **Coir ground net** with long straw mulch, grass seeding, and live stakes were used on gentler soil slopes underlain by bedrock that close to the surface. Rock was placed in transition areas between the brushlayer fill and areas where bedrock was exposed (see Fig. **4**).
- **Live fascines** were placed along crest of cut above brushlayer fill (see Fig. 5).

Fig. 3 Brushlayer fill supported on rock toe buttress

Fig. 4 Transition area treatment between brushlayer fill and exposed bedrock

Fig. 5 Placement of fascine rows at top of cut

OBSTACLES TO IMPLEMENTATION: Highway engineers were initially skeptical about ability of an earthen, brushlayer fill to support the highway cut. A number of slope stability analyses were performed with different types and combinations of hybrid retaining systems. Eventually, a satisfactory solution was devised consisting of a brushlayer earthen fill resting atop a rock toe buttress.

PERFORMANCE EVALUATION: The entire cut slope is now stable, well vegetated, and blends in with the natural surroundings (see Fig. 6). A few minor problems occurred initially that were repaired without difficulty. Rilling through the brushlayers was observed in places due to surface runoff from the crest of the slope. This problem was eliminated by constructing a brow ditch along the crest that intercepted runoff from above. Some cracking, of no consequence, was observed at the top and back of the brushlayer fill which was caused by settlement—as opposed to mass movement. This settlement was probably exacerbated by the presence of some ice in the earthen fill during construction.

The rooting and establishment of live cuttings in the brushlayer fill and fascines were quite satisfactory. An exception occurred near the base of the brushlayer fill where alder was used that was not selected and handled with the same degree of care given to the willow cuttings used in the rest of the fill. Even so the brushlayers have served their intended function and purpose. The brushlayers have provided an opportunity for native vegetation to invade and establish on the slope (see Fig. 7). As a result, the process of plant succession is well underway and, after 3 years, the project site had already assumed a natural and pleasing appearance.

Fig. 6 Brushlayer fill 2 years after construction

Fig. 7 Establishment of native volunteer vegetation in brushlayers

BENEFITS AND LESSONS LEARNED:

1. Some plant species perform better than others; preconstruction trials may be necessary to identify best performing plant material.

2. Conventional geotechnical slope stability analyses can be adapted to evaluate the stability of biotechnical slope stability installations.

3. Conventional slope stabilization techniques, e.g., a rock toe buttress, can be successfully combined with soil bioengineering techniques, e.g., brush-layering.

4. Be careful when using frozen soils in a brush layer installation—thawing and subsequent settlement can produce the appearance of a slope failure.

ACKNOWLEDGMENT

This case study is derived from material appearing in the book **Biotechnical and Soil Bioengineering Slope Stabilization** coauthored by D.H. Gray and R. Sotir, published by John Wiley and Sons, 2007. This derivative material is reproduced with permission of John Wiley & Sons, Inc.

ADDITIONAL READING:

Gray, D.H. and Sotir, R. (1992). Biotechnical stabilization of a highway cut. *Journ. of Geotechnical Engineering* (ASCE) **118** (GT9): 1395-1409.

Gray, D.H. and Sotir, R. (1992). Biotechnical Stabilization of Cut and Fill Slopes. Proceedings, ASCE-GT Specialty Conference on Embankments & Slopes, *Geotech. Spec. Publ.* #31, Vol. 2: pp. 1395-1410, July, 1992.

Gray, D.H. and Sotir, R. (1996). Biotechnical and Soil Bioengineering Slope Stabilization. John Wiley and Sons, Hoboken, NJ

PROJECT #23: BUFFALO BAYOU

PROJECT TYPE: **River bank repair and stabilization**

PROJECT SCALE: Medium

CLIENT/OWNER: Riparian homeowners

TECHNIQUES EMPLOYED: **Vegetated MSE, Chimney drain, Live fascines, Live staking, Bank spur**

GEOGRAPHIC LOCATION: Buffalo Bayou Memorial Drive Houston, TX

Fig. 1 Live fascine staked in place

GEOMORPHIC SETTING: High, steep riverbank next to bayou that is subject to frequent flooding and partial inundation from controlled releases upstream.

SITE CONDITIONS AND PROBLEMS: The combination of natural flooding and controlled releases resulted in the abrupt rise and fall of water level in the bayou in addition to prolonged periods of high water. These hydrologic conditions, combined with sandy and silty soils with little cohesion, resulted in widespread erosion and stream bank failure. The instability of over-steepened slopes is aggravated by the presence of fine sands and seepage of water from the bank face between 200 and 2,000 gallons (0.75–7.5 cu.m) per day.

Bank return flows seepage during drawdown resulted in exfiltration or seepage erosion. The stream bank was actively receding along a reach of approximately 280 ft (84 m) long and 25–35 ft (7.5–10.5 m) high on outside bend. This recession resulted in a loss of approximately 15 ft of land at top of bank and creation of near vertical scarps as a result of mass slope failures and seepage erosion (see Fig. 2).

W. Goldsmith et al., *Bioengineering Case Studies: Sustainable Stream Bank and Slope Stabilization*,
DOI 10.1007/978-1-4614-7996-3_24, © Springer Science+Business Media New York 2014

Fig. 2 Failed, eroded stream bank prior to treatment

TREATMENT OBJECTIVES AND CONSIDERATIONS: Previous treatments, consisting of bank armoring with stacked cement filled bags tied together with steel reinforcing bars, were not only unsightly but also ineffective in preventing slope failures at this site. Successful treatment required redirecting the stream away from the foot of the bank, rebuilding and securing a toe berm, and stabilization of the slope face. The interception and diversion of exfiltrating seepage from the bank was also essential to successful stabilization. In addition to these functional considerations, visual and esthetic concerns dictated a treatment that blended with the natural riparian surroundings.

A multipronged approach was adopted to deal with these various site conditions and requirements. A relatively steep fill or buttress slope with a grade of 0.5:1 (H:V) was reconstructed upon a foundation of wrapped concrete rubble that was installed in a 7-ft (2 m) deep toe-trench. The fill was constructed in 2-ft (61-cm) lifts wrapped with a synthetic geogrid and burlap. Live brush layers approximately 6 in. thick and long enough to extend from the undisturbed soil at the bank and project several feet beyond the reconstructed fill face were placed on each wrapped soil layer.

Because continued seepage and saturation of the fill would have greatly reduced the slope factor of safety, it was necessary to install chimney drains at the back of the VMSE buttress that intercepted and directed seepage downward to a rocked trench. Water collected in the trench drain in turn was discharged to the bayou via pipes located at the upstream and downstream ends of the project. A short live boom was installed at the upstream end to protect a sewer outfall and to keep the low-flow thalweg (the deepest part of the steam) away from the bank.

TREATMENTS SELECTED:

- **Vegetated MSE** buttress fill with a grade of 0.5:1 (H:V). The earthen fill was reinforced with both polymeric geogrids and live cut branches. This buttress fill was constructed upon a foundation of wrapped concrete rubble placed in a 7-ft (2 m) deep toe-trench.

- **Chimney drains** consisting of vertical strips (geodrains) were placed at the back between the VMSE and natural bank (see Fig. 3). The strip drains were installed on 5-ft (1.5 m) centers; they intercepted and collected the exfiltrating seepage and conducted it to a rocked toe-trench beneath the VMSE buttress.

- **Live fascines** were placed on a midslope bench of the VMSE reinforced fill and at the base (see Fig. 1 and Fig. 4)

- **Live staking** was used on the edges, sides, and selected areas of the project site.

- **Bank spur** (a rock groin capped with a prism of earth reinforced with criss-crossing fascines) was used to redirect stream flow away from bank.

Fig. 3 Vertical chimney drains placed at back of VMSE fill

Fig. 4 Live fascines placed on a midslope bench of the VMSE

OBSTACLES TO IMPLEMENTATION: Some skepticism voiced initially about ability of VMSE buttress to support a stream bank and protect it against stream scour and erosion. Needed to convince everyone about the destabilizing effect of ex-filtrating seepage during drawdown phase in the bayou and to stress importance of installing chimney drains at the back of the VMSE buttress to intercept and redirect this seepage.

PERFORMANCE EVALUATION: The site has remained stable since construction, and the soil bioengineering installation has developed into a dense riparian buffer as native and naturalized species of herbaceous and woody vegetation invade the site and become established. The site survived a major bankfull flood during the second year after construction. The supple stems of woody vegetation on the bank bent over during the flood and protected the bank against scour erosion. Minor maintenance and pruning may be required periodically to keep the bank from becoming to overgrow and to remove large trees from the top or midslope portions of the bank.

<u>LESSONS LEARNED</u>:

1. A well designed vegetated mechanically stabilized (VMSE) buttress can be used to support and protect a stream bank against both scour erosion and slope failure.

2. Where there is substantial bank seepage, a chimney drain is required behind a brush layer or vegetated (VMSE) installation to provide adequate drainage and prevent saturation or buildup of excess pore water pressures within the installation.

3. Adequate toe protection in the form of rock armor and/or flow redirection in the form of spurs or rock vanes is sometimes required in conjunction with protection on the bank slope or face.

ACKNOWLEDGMENT

This case study is derived from material appearing in the book **Biotechnical and Soil Bioengineering Slope Stabilization** co-authored by D.H. Gray and R. Sotir, published by John Wiley and Sons, 2007. This derivative material is reproduced with permission of John Wiley & Sons, Inc.

ADDITIONAL READING:

Nunnally, N.R. and Sotir, R. (1994). Soil bioengineering for streambank protection. *Erosion Control* **1**(4): 39-44.

Sotir, R. (1995). Soil bioengineering experiences in North America. In *Vegetation and Slopes*, D.H. Barker (Ed.), Thomas Telford: London, pp.190-201

Gray, D.H. and Sotir, R. (1996). Biotechnical and Soil Bioengineering Slope Stabilization. John Wiley and Sons, Hoboken, N.J

PROJECT #24: LITTLE TOPASHAW

PROJECT TYPE: **River bank repair and stabilization**

PROJECT SCALE: Medium

CLIENT/OWNER: USDA-NRCS and Chickasaw County Conservation Dis.

TECHNIQUES EMPLOYED: **<u>LW Structures, Willow posts and poles</u>**

GEOGRAPHIC LOCATION: Topashaw Creek, Yalobusha River watershed, North Central, MS

Photo courtesy of F.D. Shields

Fig. 1 Large wood structure

GEOMORPHIC SETTING: Fourth-order stream. Floodplain stratigraphy is characterized by dispersive silt and clay soils underlain by sand that overlies consolidated cohesive material. Sandy deposits are often found along the bank toe. The channel is tortuous, with an average sinuosity of 2.1, an average width of about 115 ft, and an average depth of 13 ft.

SITE CONDITIONS AND PROBLEMS: The most striking characteristics of the study reach were its flashy hydrology and physical instability. Stream flow reflected previous upstream straightening, incision, and the absence of floodplain storage (Shields et al. 2008). High flow events tended to be extremely brief (<30 h) and frequent, with maximum depths of approximately 3 m and occasional velocities as great as 3 m/s. Available evidence suggests mean channel width had increased by a factor of 4 to 5 since 1955. Geomorphic evaluation performed prior to construction indicated that the upstream and middle reaches were still degrading. In general, concave banks on the outside of meander bends were caving, and sand is accreting on large point bars opposite failing banks. Outside of bends, eroding banks frequently invade adjacent cultivated fields (See Fig. 2), while inside bends and abandoned sloughs were vegetated with a diverse mixture of hardwood trees and associated species.

DOI 10.1007/978-1-4614-7996-3_25,

(a) *Photo courtesy of F.D. Shields* **(b)** *Photo courtesy of F.D. Shields*

Fig. 2 Physical instabilities along Little Topashaw Creek prior to rehabilitation: (**a**) Bank erosion and (**b**) Riparian gully incision

TREATMENT OBJECTIVES AND CONSIDERATIONS: Structural measures for stream corridor rehabilitation must be selected and designed to harmonize with the dominant geomorphic processes. For ecosystem rehabilitation, they must address the major factors inhibiting natural recovery. Accordingly, this project was designed to accelerate evolution of the existing system toward a sinuous two-stage channel with wooded berms. Bank stabilization structures made from large wood (LW) instead of stone were placed along the toe of eroding banks. The large wood structures (LWS) were designed to resist displacement by inter locking, keying-in to banks, anchoring, and inducing sediment deposition. The LWS were intended to accrete and retain sediment and organic matter introduced both from adjacent mass wasting and material transported into the reach from upstream. Crest elevations were set high enough to stabilize existing, near-vertical banks that were failing by mass wasting. This was accomplished by building a berm at the toe and preventing episodic cleanout of failure blocks.

TREATMENTS SELECTED:

1. **LWS.** About 1,500 m (5,000 ft) of eroding banks were selected for LWS protection. LWS were constructed using either woody debris (~10 %) or living trees (~90 %) harvested from designated areas, including the channel. Living trees were larger than 200 mm (8 in.) diameter at breast height (DBH). Living trees were harvested by grubbing in order to retain root balls and crowns intact. LWS were constructed by stacking trees. The members running across the flow direction ("key members") were ~9 m (30 ft) long and were keyed into the bank toes when bank slopes were gradual enough to permit key trench excavation. An average of 16 trees was used per LWS. Structures were spaced to create nonuniformity, which is valuable for physical habitat recovery, but were aligned to enhance log stability and sediment deposition. In general, structures extended about 15 m (50 ft) in the stream wise direction, about 5 m (16.5 ft) transverse to the stream, and were spaced about 14 m (46 ft) apart (See Fig. 1 and Fig. 3). About one LWS was

constructed to protect each 20 m (65 ft) of eroding bank, which represented an order-of-magnitude increase in LW loading.

2. **Willow poles and posts.** Approximately, 4,000 willow *(Salix nigra)* cuttings were planted on point bars and in sediment deposits adjacent to selected LW structures using a water jetting technique (See Fig. 4). Willow cuttings were soaked for 2–3 weeks prior to planting to improve transplant success. A subset of the willows was planted in square plots selected to represent a range of soil textures and vertical distances from the base flow channel to relate willow growth and survival to environmental conditions.
3. **Native plantings.** Selected grasses were installed to rapidly stabilize accreted deposits and gullies formed by runoff passing over the tops of banks. Fourteen edge-of-field gullies were identified and surveyed for this treatment along both banks of the study reach. Six of these were designated for treatment with switchgrass hedges, leaving eight as untreated controls.

(a) **(b)** *Photo courtesy of F.D. Shiel*

Fig. 3 Large wood (LW) structure deployed at Little Topashaw Creek. (**a**) Design plan and profile. (**b**) LW structure 5 months after construction

(a) *Photo courtesy of F.D. Shields* **(b)** *Photo courtesy of F.D. Shields*

Fig. 4 Willow cuttings planted at Little Topashaw Creek. (**a**) Willows being planted using water jetting. (**b**) Cuttings 4 months later showing effects of herbivory

OBSTACLES TO IMPLEMENTATION: Some skepticism was voiced initially by client about ability biotechnical methods to protect stream banks. Habitat materials available for LWS construction were limited to LWD presently in the channel and trees growing in patchy stands on the floodplain. No clearing was permitted within 10 m (33 ft) of top bank. The finished project consisted of 72 structures built with about 1,200 trees, but these were obtained by clearing only about 3.4 ha (8.4 acres). Cleared areas were primarily zones such as fencerows and ditches that landowners wanted cleared for cultivation.

PERFORMANCE EVALUATION: Initially, the large wood (LW) structures were quite effective at modifying flow fields, trapping sediments, and stabilizing steep banks. During the high flow season prior to construction, measured top bank retreat rates were 1.1–1.7 m, but these declined to 0.44–0.66 m during the first year following construction. Sixty-eight of the 72 structures survived the first high flow season, and almost all accreted significant volumes of sediment.

In the long run the physical instability of the study reach was only slightly ameliorated by rehabilitation, and flashy stream hydrology was not addressed by reach-scale measures. Although the reach-scale measures of LW structures and willow planting were initially successful, at the end of the second year, 31 % of the LW structures had been destroyed and 22 % were damaged, and annual bank retreat rates had increased from 0.44 to 1.6 m. Only one-third of the 72 LWDS were undamaged 3 years after construction, while one-third had been destroyed. Willow survival depended upon location on the bank and was affected significantly by herbivory. After two growing seasons, the mean willow survival was only ~50 % in study plots that were >0.5 m above the stream at base flow, whereas survival in lower plots averaged 86 %. After 5 years, willow survival was less than 10 %.

LESSONS LEARNED:

1. Studies of degraded streams across the region have shown that habitat diversity, invertebrate species richness and abundance, and fish species richness are associated in a positive fashion with the addition of LW structures. However, in the case on Little Topashaw Creek, these improvements were temporary.

2. The successful use of plant materials and LW structures for stream corridor habitat rehabilitation is limited by geomorphic processes that act at the landscape scale (Shields et al. 2008). In the case of Little Topashaw Creek only traces of the LW structures and willow plantings remained after 5 years.

ACKNOWLEDGMENT

Helpful information about this project was provided by Dr. F.D Shields, Jr. He permitted the reprinting of the photos in the case study. His assistance is greatly appreciated and acknowledged.

ADDITIONAL READING:

NCHRP (2004). Environmentally Sensitive Channel and Streambank Protection Measures, NCHRP Report 544. Guidance manual prepared for the Natl. Coop. Hwy. Research Program, Washington, DC.

Shields, Jr., F.D., S.S. Knight, and Stofleth, J.M. (2006). Large wood addition for aquatic habitat rehabilitation in an incised, sand-bed stream, Little Topashaw Creek, Mississippi. *River Research Applications,* Publ. online in Wiley InterScience. U.S. Govt. work in the public domain in the USA

Shields, F. D., Jr., Morin, N. and Cooper, C. M. (2004). Large woody debris structures for sand bed channels. *Journal of Hydraulic Engineering*. 130 (3):208-217.

Shields, Jr. F.D., S. Reza Pezeshki, G.V. Wilson, and S.M. Dabney (2008). Rehabilitation of an incised stream using plant materials: the dominance of geomorphic processes. *Ecology and Society* **13**(2):54

PROJECT #25: NEW CONCORD

PROJECT TYPE: **Hwy embankment fill stabilization**

PROJECT SCALE: Medium

CLIENT/OWNER: Ohio Department of Transportation

TECHNIQUES EMPLOYED: **Live willow poles**

GEOGRAPHIC LOCATION: Intersection of 1-70 and SR83 near New Concord, OH

(photo courtesy of Prof. T.H. Wu)

Fig. 1 Root development around an exhumed willow pole

GEOMORPHIC SETTING: Slope along major interstate highway and state road underlain by either stiff or compacted clay both of which tend to lose strength with age. The layer near the surface is subject to wet-dry and freeze–thaw cycles; the cohesive strength gradually approaches zero. During the wet season, the ground becomes saturated.

SITE CONDITIONS AND PROBLEMS: Upon saturation and under vertical seepage, a 2H:lV slope would have a safety factor close to l.0. Local slips develop because of nonuniform soil and seepage conditions. Such slips are not usually noticed by maintenance crews. Over a period of several years, however, these slips grow, coalesce, and extend to greater depths. The condition becomes critical when the upper limit of a failure reaches the shoulder. To protect the roadway, the conventional repair measure is to excavate the soft material and replace it with compacted fill, which is fairly costly. Then the cycle is repeated. The service life of such slopes ranges between 10 and 20 years, depending on the site conditions.

W. Goldsmith et al., *Bioengineering Case Studies: Sustainable Stream Bank and Slope Stabilization*,
DOI 10.1007/978-1-4614-7996-3_26, © Springer Science+Business Media New York 2014

TREATMENT OBJECTIVES AND CONSIDERATIONS: The main goal was to stabilize shallow slope failures. To achieve this goal live poles were installed in the zones with shallow slips. Willowpoles were installed in adjacent sections, which had no slips, as a preventive measure. The lateral resistance of the poles contributes to shearing resistance along the sliding surface and increases the safety factor. In addition, the root systems of established poles can contribute an additional component to the shearing resistance. On the other hand, if the poles die and decay over time, their contribution to stability will also disappear. Thus, the long-term vitality of the poles is critical.

TREATMENTS SELECTED:

Live willow poles: The poles consisted of stems cut from willow and poplar trees located within 20 km from the sites, during early spring (March–April) and prior to sprouting of leaves. The poles were trimmed to the required dimensions. The poles increase resistance to shallow slope failures by acting as shear pins. Adventitious root development around the poles (see Fig. 1) provides some fiber reinforcement as well. The installation at New Concord was delayed until late May 2005. It became necessary, therefore, to store the harvested stems in refrigeration (at 2 °C) as has been done in the UK until installation. In the present study, live poles approximately 2 m long and 25–50 mm in diameter were installed vertically in a grid pattern approximately 1 m on center (see Fig. 2). All poles were placed in pre-made 75 mm diameter holes. Two installation methods were used. Method A is considered to be the best method and is also the most labor intensive. In this method, each pole is cut to a bevel at the bottom, the top is wrapped with wire to prevent splitting, the pole is driven 150 mm into the bottom of the hole with a mallet, and then the top is cut off, rewired, and painted with a protective sealant. For Method B, which is simpler and less labor intensive, the poles are simply dropped into pre-made holes, backfilled, and painted with a protective sealant.

OBSTACLES TO IMPLEMENTATION: The main obstacle to successful implementation appears to be soil moisture deficiency during the critical establish ment phase. The low survival rates for the first installa tion (May 05) at New Concord (50 %) were attributed to the late date of installation, which was followed by a period of hot weather. The principal limitations of willow pole plantings are constraints on construction time and the procedures for installing plant materials, which are critical but unfamiliar to many civil engineers and contractors. Also, the survival of poles requires additional and maintenance beyond what is typically necessary for conventional stabilization repairs.

(photo courtesy of Prof. T.H. Wu)

Fig. 2 Willow pole array installed on highway embankment slope

PERFORMANCE EVALUATION: Slope performance before and after stabilization was monitored using inclinometers, piezometers, tensio meters, and moisture blocks. Load tests on the poles were conducted to measure vertical and lateral resistances. To monitor the shallow slips, copper tubes and plastic tubes 10 mm in diameter were placed in pre-drilled holes. The positions of their tops were monitored. Excavations were then made on one side to expose the tubes. This showed that the movements were limited to the top 0.6 m. In addition, a row of stakes were placed along the slope, and their movements were monitored relative to fixed benchmarks. The measurements showed that shallow movements of 5 cm occurred during the winter of 2005–2006 at New Concord. Most of the dead poles at New Concord were replaced with willow poles during the following spring (Feb 06). These poles were installed shortly after harvest and had a good survival rate (90 %). After the repairs at New Concord, the movements were minimal during the following wet season, although soil moisture remained high and near saturation.

COST ANALYSIS

Detailed cost information was recorded on this project because the comparative cost of bioengineering remediation vs. conventional techniques was of particular interest to the Ohio DOT. Cost and other detailed analyses can be found in the technical paper by Kokesh et al. (2013). According to these authors, the total cost of construction, including slope regrading, was \$25,000 and the total area of construction was 240 m^2, which yields an average unit cost of \$100/m^2. A production level of approximately 25 fully installed poles per day was achieved. The cost of the soil-bioengineering repair was 30 % lower compared to conventional techniques. This cost can be expected to decrease even further as local contractors become familiar with soil bioengineering techniques.

Construction and installation in the field was performed primarily by students. Accordingly, the total cost for this project may be lower than that for a typical contractor. It's important to note that had this slope not been repaired, a large scale failure could have been expected in the future. Such a failure would require much more costly reconstruction than the remediation described in this case study. Therefore, soil-bioengineering was a cost-effective approach as a proactive maintenance strategy for this project.

BENEFITS AND LESSONS LEARNED:

1. Results obtained to date indicate that live poles, if they can be established, can be effective for stabilization of shallow slides at depths of approximately 1.5 m or less.

2. Potential benefits of soil-bioengineering are reduced time and cost of repair and minimum or zero interruption of traffic.

3. Techniques and protocols should be adopted to aid establishment of live willow poles during hot weather conditions, e.g., use of DriWater polymer granules, when soils tend to dry out

ACKNOWLEDGMENT

Information about this project was provided by Professor T. H. Wu, Dept. of Civil Engineering, Ohio State University, who directed the project. His assistance is greatly appreciated and acknowledged.

ADDITIONAL READING:

Steele, D. P., Barker, D., et al. (2004). "The use of live willow poles for stabilizing highway slopes." Report TRL619, TRL Ltd., Berks

Wu, T.H., Fox, P.J., Barker, D.H. et al (2008). Soil bioengineering for slope stabilization. Paper presented at GeoCongress 2008.

Kokesh, C.M., Trenner, B.R., Wu, T.H., and Fox, P.J. (2013). Soil Bioengineering for Slope Stabilization: A Case History. Paper submitted for publication in the Journ of Geotechnical and GeoEnvironmental Engineering, ASCE.

PROJECT #26: WATER PURIFICATION FACILITY AND PARK

PROJECT TYPE: **Shoreline Stabilization**

CLIENT/OWNER: South Central Connecticut Regional Water Authority

PROJECT SCALE: Medium

TECHNIQUES EMPLOYED: Coconut Fiber Rolls Live Brushlayering, Live Staking, Green Roof, Bioswales

(photo courtesy of Elizabeth Felicella)

Fig. 1 Treatment facility park site

GEOGRAPHIC LOCATION: New Haven and Hamden, CT

GEOMORPHIC SETTING: The site is located on a sensitive tributary of Long Island Sound. The site, which lies within the upper reaches of an estuary, is underlain by the Connecticut Valley Arkosic Sandstone. It can be viewed from a nearby overlook atop a basaltic sill. A parcel of land that includes the site, a dam, and reservoir were deeded by Eli Whitney to the people of region to serve its water supply needs (Fig. 1).

SITE CONDITIONS AND PROBLEMS: Many objections were voiced about siting a large, complex, modern treatment facility on a sensitive, constrained, and highly regulated site. Multi-stakeholder satisfaction required addressing these concerns by incorporating educational benefits and by identifying opportunities and methods to integrate building, landscape, utility, and permitting challenges. A site formerly occupied by an obsolete water treatment plant was reused, and demolished building materials were incorporated into the earthworks. All excavated earth and land clearing debris were salvaged and reused on the site; new building materials were selected in keeping with sustainable design.

W. Goldsmith et al., *Bioengineering Case Studies: Sustainable Stream Bank and Slope Stabilization*,
DOI 10.1007/978-1-4614-7996-3_27, © Springer Science+Business Media New York 2014

TREATMENT OBJECTIVES AND CONSIDERATIONS

The project was intended to show how industrial facility development can employ low-impact and sustainable measures compatible with water supply management as well as recreation, habitat, and educational access. Best management practices were adopted and integrated together for the entire site, including a green roof, bioswales, treatment pond, and wetlands. The project serves as a state-of-the-art model for effective watershed stewardship. Water treatment processes were selected for efficiency, relying on gravity flow instead of pumping. Energy systems were based upon geothermal loops. Storm water handling was achieved through the use of bioengineered measures that incorporated vegetation rather than artificial catch basins and pipes. As a result, off-site impacts such as energy demand, waste handling, and runoff were responsibly addressed.

TREATMENTS SELECTED:

- **Green Roof.** A lightweight engineered growing substrate supports alpine type vegetation and allows the building to absorb rainfall and slowly release moisture via evapotranspiration (see Fig. 2). As a result the building benefits from natural cooling and does not impact the surrounding watershed with excessive runoff. Visual and habitat benefits are enjoyed by neighbors and butterflies. The green roof was designed to accommodate access to allow people to view down into the treatment facility and observe the various stages of the purification process through porthole skylights.

Fig. 2 Green roof allows building footprint to function naturally

(photo courtesy of Elizabeth Felicella)
Fig. 3 Bioswales conveying water to ponds and wetlands

- **Bioswales.** Collected water from pathways, lawns and native wildflower meadows, and vegetated drainage swales is carried into a multipurpose treatment pond and wetland complex. Vegetation stabilizes the soil and provides nutrient uptake and habitat values (see Fig. 3).

- **Coconut Fiber Rolls.** Coconut Fiber Rolls prevented soil erosion along pond shores and served as stable planting substrate for herbaceous vegetation capable of withstanding constant saturation and frequent inundation.

- **Live Brushlayering.** For pond shoreline areas designed with steeper slopes, layers of dormant cuttings of dogwood, elderberry, and willow were embedded into soil fills, allowing precise and graceful grading, while offering beauty and habitat values.

- **Live staking.** Where pond shorelines were not as steep, stakes of woody shrub species were added to help foster a healthy buffer of vegetation around the pond and wetland features.

OBSTACLES TO IMPLEMENTATION: Initial concerns of regulators and others were addressed satisfactorily by the development of an innovative, creative, and sustainable design. The next hurdle was to conduct and provide a convincing benefit–cost analysis. The design team worked closely to characterize the long-term cost savings from a natural systems approach, and the rigorous of use of recycling demolished materials. Success required evaluation beyond initial construction costs alone. Life cycle operational savings including reduced energy demand and use of passive systems also had to be considered.

PERFORMANCE EVALUATION: Surface runoff drains to an unlined pond where it percolates into the underlying the soil. The pond also treats back-flush water from the facility's filter system and absorbs overflow from the finished drinking water tank. The facility sets a new standard for interaction among municipal infrastructure, community space, and the surrounding ecosystem (see Fig. 4). The new facility demonstrates that low-impact site development can be compatible with recreation, habitat, and educational access. The green roof was designed to accommodate access so that people can observe the various stages of the purification process. The project has received numerous awards including the 2010 ASLA General Design Honor Award and the 2007 AIA Committee on the Environment Top Ten Green Award.

BENEFITS AND LESSONS LEARNED

Using techniques adapted from restoration ecology and bioengineering, the landscaped site creates a microcosm of the surrounding regional watershed, from mountain source to reservoir (refer to Fig. 4). By integrating bioengineering measures, the entire site, including green roof, vegetated swales, and treatment wetlands, serves as a state-of-the-art model for effective watershed stewardship. In 2010, the American Society of Landscape Architects Professional Awards chose the South Central Connecticut Regional Water Authority's Water Purification Plant for its General Design Honor Award.

(photo courtesy of Alex MacLean)

Fig. 4 Building and water features linked for beauty and function

ADDITIONAL READING:

Kundert, Robert (2008). A Transition from NIMBY to Amenity: Creative sustainable approach to Connecticut public works project wins over neighbors and captures 2007 Visionary Project Award. Sustainable Land Development Today

Stowe, Stacey(2005). Connecticut Town Helps Create an Architectural Anomaly: An Appealing Water Plant. New York Times, December 4, 2005

PROJECT #27: WALDEN POND

PROJECT TYPE: **Shoreline Restoration**

PROJECT SCALE: **Large**

CLIENT/OWNER: Massachusetts Department of Conservation and Recreation

TECHNIQUES EMPLOYED: **Vegetated MSE Blueberry Sod Live Brushlayering, Live Staking**

GEOGRAPHIC LOCATION: Walden Pond Concord, MA

Fig. 1 View of Walden Pond

GEOMORPHIC SETTING: Walden Pond is a 62-acre glacial kettle-hole pond in a Massachusetts state park and is a National Historic Landmark. Walden Pond was the setting of Henry David Thoreau's classic, "Walden," which has been read by thousands for over a century. Thoreau, a poet, philosopher, and naturalist, is considered to have begun the American environmental movement. The pond has attracted large crowds of bathers since Victorian times (Fig. 1).

SITE CONDITIONS AND PROBLEMS: Water quality and vegetative richness were diminished by upland erosion caused by uncontrolled circulation and use. Decades of heavy foot traffic compacted soils, degraded the health of forest under-story, triggered erosion, and diminished the visual quality and integrity of the Pond's shorelines. Previous efforts to rehabilitate the Pond's perimeter path and to stabilize the banks were ill-suited to the site's natural appearance; furthermore, the path was in poor condition, unsafe, and needed repair. A hodgepodge of granite curbstones, salvaged railroad ties, and other materials had been used but were not functional or attractive. A later effort in the 1980s to use boulders, interplantings, and erosion control blankets was inconsistent with the geomorphic setting and Thoreau's historic description of the pond.

W. Goldsmith et al., *Bioengineering Case Studies: Sustainable Stream Bank and Slope Stabilization*,
DOI 10.1007/978-1-4614-7996-3_28,

TREATMENT OBJECTIVES AND CONSIDERATIONS: A vocal stakeholder group consisting of local frequent visitors, environmentalists, and perceptive historians wished to restore the appearance of Walden just as Thoreau described in his writings, right down to plant species. State park operators were required to provide ADA compliant accessible trails throughout the park, including areas that had become entirely impassible and unsafe. The goal was to provide a new trail alignment that would preserve existing healthy forest and shoreline cover, repair severely eroded banks, intercept runoff, and provide controlled access for people at convenient locations. The latter were chosen to minimize impacts to the viewshed. This complex and ambitious process also required use of tree, shrub, and groundcover species not commonly used in bioengineering applications, in order to obtain historically and visually compatible results.

TREATMENTS SELECTED:

- **Vegetated MSE.** A mechanically stabilized earth structure was placed along portions of the shoreline to provide a foundation for the newly aligned path system (see Fig. 2). Lifts of soil were encapsulated in high tensile strength geogrids in deep, buried positions, whereas a heavy coir mesh fabric was used in exposed, above-ground locations. Vegetation native to the area was inserted into the earthen structure in the form of live stakes and live brush layers (between lifts) as well as contract-grown tubelings of species that were not feasible to propagate from dormant cuttings.

- **Blueberry Sod.** Steep and bare cut banks were exposed upslope from the newly located path. Lowbush blueberry and other species, e.g., commercially available Blueberry Sod, were pegged in place to match the surrounding understory (see Fig. 3). This treatment allowed disturbed areas to rapidly achieve the same level of plant cover as the most stable areas of the forest floor. The sod itself contained diverse plants representative of New England oak-pine forest understory that had been salvaged from development sites.

- **Live Brush Layering.** In addition to inserting Brushlayers between MSE lifts, they were also used in areas where deep gullies needed repair on upper slopes above the pathways. Species used for this purpose included willow, dogwood, raspberry, blackberry, elderberry, and alder.

- **Live Staking.** Willow, alder, buttonbush, and dogwood cuttings were inserted as live stakes through the coir mesh and into the MSE structures. They were also used to revegetate other bare spots at the treatment site.

Fig. 2 Path atop Vegetated MSE slope along pond shore

Fig. 3 Blueberry sod on cut banks

OBSTACLES TO IMPLEMENTATION

This site is a national historic treasure that evoked emotional and passionate responses to proposed changes from locally involved stakeholders. Restoring the project area to preserve a natural appearance while maintaining public access and utilizing only species mentioned by Thoreau in his descriptions of Walden Pond presented a unique challenge. A major contributor to bank destabilization was off-path hiking by park users; deterring this action, especially during the vegetation establishment period, was of particular concern. Reestablishment of a healthy vegetative cover was not possible without measures to control foot traffic. Accordingly, temporary fencing was used to exclude park goers from sensitive new plantings during construction and establishment. However, public users vocally resisted the idea of fencing.

BENEFITS AND LESSONS LEARNED

By choosing simple, low cost range fencing, park visitors were excluded from sensitive areas. At the same time this type of fence blended into the surroundings unobtrusively and was essentially invisible from a distance. People ultimately became accustomed to it and recognized its effectiveness at preventing damage. As a result the public requested that the fence be kept in place permanently. Blueberry sod was a novel measure that suited the site very well and added a high level of visual and historical integrity to the restoration effort.

ADDITIONAL READING:

Goldsmith, Wendi. (2001). "Improving the Nature of Walden Pond." *IECA News* Nov.-Dec. 1998:

PROJECT #28: HEARTHSTONE QUARRY BROOK

PROJECT TYPE: **Channel stabilization**

PROJECT SCALE: Medium

CLIENT/OWNER: Department of Public Works, City of Chicopee

TECHNIQUES EMPLOYED: **Cross Vanes, Vegetated Gabion Baskets, Live Staking, Live Fascines, Coconut Fiber Rolls, Live Brushlayering**

GEOGRAPHIC LOCATION: Chicopee, MA

Fig. 1 Cross vanes

GEOMORPHIC SETTING: The Hearthstone Quarry Brook is a headwater stream that runs through a dense residential area and receives storm water run-off from area roads, highways, and parking lots in addition to significant inputs from the Massachusetts Turnpike. Soils are extremely fragile, noncohesive glacial outwash sands which are incised by deep ravines including the stream corridor itself and various small intermittent tributaries. Base flows are relatively constant due to highly permeable subsoils that groundwater discharge into the stream (Fig. 1).

SITE CONDITIONS AND PROBLEMS: Planned road expansion threatened to generate even more storm water flow to the brook, forcing the city to address problems caused by severe channel incision in addition to general erosion which would jeopardize roads, outbuildings, homes, and sewer lines. Hence, the planned road expansion provided an opportunity for comprehensive treatment of chronic but unaddressed problems. These included massive channel instability that could trigger steep bank failures and downstream sedimentation in the brook. Multiple homes abutted the Brook; in some cases garden sheds, garages, and other structures had fallen in due to collapse of steep banks.

W. Goldsmith et al., *Bioengineering Case Studies: Sustainable Stream Bank and Slope Stabilization*,
DOI 10.1007/978-1-4614-7996-3_29,

TREATMENT OBJECTIVES AND CONSIDERATION:

Major considerations for this project included implementing treatment methods that would use the stream corridor itself for access. This was required to avoid disturbance to private property and steep fragile banks along the stream. Additionally, the site was heavily shaded because of the incised ravine conditions, as well as dense surrounding forest cover. Therefore, special plant choices were required to ensure channel stability and control erosion while establishing vegetation which would thrive without full sun. Major restoration goals included creating a corridor for use by urban wildlife, restoring riparian buffers to aid in maintaining water quality, correcting channel incision to improve floodplain connection, and restoring natural bank vegetation and woody debris.

TREATMENTS SELECTED:

- **Vegetated Gabion Basket check dams.** Several gabion check dams were used to provide grade control and elevate the stream profile to reconnect to the established floodplain. Natural log dam structures were also used to supplement grade control and enhance aquatic habitat (see Fig. 2). Various species of vegetation were planted into the gabions to increase riparian functions.

- **Coconut Fiber Rolls.** To define bank location and dimensions in the non-cohesive sandy channel materials, a nearly continuous row of Coconut Fiber Rolls was placed at the toe of both banks. In some narrow reaches with steep banks, rolls were stacked too high and laced together to provide an integrated soft armor protection to the bank. Shade-tolerant sedge species were planted into the rolls after placement.

Fig. 2 Gabion check dam used to provide grade control and log vane for aquatic habitat

- **Cross vanes.** Existing large woody debris on site was salvaged to create cross vanes as small grade management structures, as well as single vane energy dissipation and habitat enhancement structures.

- **Live Staking.** Plant materials selected for the site included various dogwood, viburnum, and willow species known to perform well in various levels of shade. Stakes were applied to the upper banks, as well as around gabion and log structures where additional strength from vegetation helped provide a stable transition between soil and structure.

- **Live Fascines.** To prevent rill formation on steep slopes, fascines made from dogwood, willow, and viburnum species were installed in shallow trenches and anchored onto the banks.

- **Live Brushlayers.** Where groundwater seepage was present and/or where banks were most steep and tall, live brush was deeply embedded into soil as fill was placed. In locations with severe seepage, brushlayers were supplemented with geogrid wraps to reinforce soil lifts.

OBSTACLES TO IMPLEMENTATION

Poor access due to the confined incised valley (see Fig. 3), adjacent residential property, and saturated soils made the stream channel itself the only viable means of access for equipment and materials. Construction took place in the fall, and bypass pumping redirected stream flow during restoration work.

Fig. 3 Vegetation emerging from planted Coconut Fiber Rolls, Log Vanes, Live Staking, and Live Brush Layering in spring

PERFORMANCE EVALUATION

Formal monitoring of projects happens all too seldom, however pre-construction and three years of post-construction monitoring in the case of the Hearthstone Quarry Brook project. The monitoring plan covered stability, habitat values, and vegetation performance. Results of this monitoring program showed that live stakes performed well with an average survival rate of 67 %, variable by habitat.

Channel stability and erosion rates were surveyed. The channel gradient was reduced; the channel profile and permanent cross sections remained stable. Only minor localized scour was apparent. Wildlife habitat was evaluated; a formerly sterile stream corridor had improved significantly. The density of woody vegetation and aerial cover improved, in spite of temporary loss due to selective tree removals (see Fig. 4). Diverse macro-invertebrates returned indicating good water quality in the streambed; small animals now used the stream corridor.

BENEFITS AND LESSONS LEARNED:

Fragile sandy soils had not previously been considered good candidates for bio-engineering treatments. In addition, shady conditions were known to present severe limitations to plant growth. The Hearthstone Quarry Brook project demonstrated how these factors could be overcome through careful design and implementation that included the use of supplemental structures.

Fig. 4 Appearance of stream channel and corridor after restoration treatment

ADDITIONAL READING:

Goldsmith, Wendi, Kirk R. Barrett, Marit Larson, and William Lattrell. "Urban Channel Restoration: Design and Monitoring: Engineering Approaches to Ecosystem Restoration," *Proceedings*, pp. 1025-1031, Reston, VA: ASCE, Wetlands Engineering and River Restoration Conference, March 22-27, 1998, Denver, Colorado.

Wang, S. Y., Eddy J. Langendoen, and F. Douglas. Shields (1997). Management of Landscapes Disturbed by Channel Incision: Stabilization, Rehabilitation, Restoration. Center for Computational Hydroscience and Engineering, University of Mississippi. 177 pp

PROJECT #29: MILL CREEK CENTER HILL LANDFILL

PROJECT TYPE: **Streambank Stabilization**

PROJECT SCALE: Large

CLIENT/OWNER: City of Cincinnati Municipal Sewer District

TECHNIQUES EMPLOYED: **Vegetated MSE, ECBS**
Willow Poles
Live Brushlayering
Live Fascines and Stakes

GEOGRAPHIC LOCATION: Lower Mill Creek Cincinnati, OH

Fig. 1 View of Mill Creek

GEOMORPHIC SETTING: The Mill Creek watershed comprises 166 square miles in the heart of Hamilton County, Ohio. Despite the fact that the watershed as a whole is very flashy, the Creek itself is relatively stable and exhibits slow rates of change, largely due to the presence of dense, tightly varved glacial lakebed clays which confine the channel bed and lower banks. The watershed consists predominantly of developed urban areas with pockets of forested land. Small areas of land under park and agricultural use still exist in the northern portion of the watershed. The lower 8 miles of the Mill Creek have been permanently modified by the US Army Corps of Engineers (USACE) through the construction of a concrete lined flood control channel.

SITE CONDITIONS AND PROBLEMS: The Center Hill Landfill in Cincinnati had been capped in 1977, and a collection system was installed to pump intercepted leachate to a sanitary sewer. However, in the 1990s significant bank erosion adjacent to the landfill had caused solid waste and portions of the leachate collection system to discharge into the Creek. The City of Cincinnati installed sections of riprap along the bank from the toe to the top as an emergency measure. When erosion threatened to flank the armored area, it became evident that a more comprehensive, long-term bank stabilization solution was necessary (Fig. 1).

W. Goldsmith et al., *Bioengineering Case Studies: Sustainable Stream Bank and Slope Stabilization*,
DOI 10.1007/978-1-4614-7996-3_30, © Springer Science+Business Media New York 2014

TREATMENT OBJECTIVES AND CONSIDERATIONS: The erosion had to be corrected to prevent leachate from contaminating the creek. A new leachate collection system also had to be installed. Additionally, tree species capable of removing large volumes of water by evapotranspiration (phreatophytic vegetation) were selected in order to intercept and reduce the amount of leachate in the ground. This in turn reduced the amount of leachate pumping and treatment each of which requires significant energy use. In order to establish vegetative cover for habitat value on the lower bank, the existing riprap was reused rather than removed. Stone was configured to provide continuous toe protection along the eroded reach, connecting into stable points up- and downstream, so flanking could not occur. Vegetation was added to increase resistance and hydraulic roughness.

TREATMENTS SELECTED:

- **Live Poles.** Using the "stinger method" to poke a hole through riprap with a strong metal probe mounted onto an excavator bucket, thick diameter dormant willow cuttings were inserted into and through the lower bank stone armor. Willows grew much better in sediment trapped among the stone than possible in the ancient dense clay subsoils on the lower bank area.
- **Vegetated MSE.** The bank was graded and stepped back at an angle of approximately 1.5:1 (H:V) to provide a temporary access platform for equipment and workers. Geogrid was placed on each bench and wrapped around lifts of fill soil (see Fig. 2).
- **Live Brushlayering.** In addition, live dormant brushlayering was inserted between successive lifts of geogrid and soil.
- **Live Fascines.** Upper bank areas were treated with fascines placed on the contour line to aid in intercepting and infiltrating any overland runoff that may develop in the future as land use changes.
- **Live Staking.** Areas on the bank were treated with dormant cuttings known as Live Stakes. Species selected for this purpose included willow, maple, alder, box elder, sycamore, dogwood, and poplar.
- **Erosion Control Blankets (ECBs).** Erosion Control Blankets were placed along the upper slopes to provide temporary erosion protection and support establishment of grass seed.

OBSTACLES TO IMPLEMENTATION

The highly regulated landfill setting, combined with the federal flood control channel, created multiple and often competing constraints and expectations related to the treatment of the project site.

Fig. 2 View of installed Vegetated MSE system

PERFORMANCE EVALUATION

Despite the novelty of the techniques in this geographic region, the project was completed on time and within budget. The regulatory community, local citizens groups, and municipal engineers were all pleased with the design and its performance (see Fig. 3). Additionally, the design offered a savings of 30 % over a different bioengineering design by others, while simultaneously being favored for its ecological integrity and stability.

BENEFITS AND LESSONS LEARNED:

After more than 5 years of highly effective performance, the community opted to expand the treated area to cover 2 miles of riverbank upstream and downstream of this site. A public greenway was incorporated to allow recreational access to what has become a stable, beautiful, and biologically diverse riparian corridor (see Fig. 4). This occurred despite the site had previously been identified as a stigmatized, unstable, and contaminated area.

(a) (b)

Fig. 3 Views of completed bank protection system: (**a**) Immediately after construction. (**b**) 3 weeks later

Fig. 4 View of the site 4 years after construction showing rapid reforestation

ADDITIONAL READING:

Barrett, K; W. Goldsmith, M. Silva (2006). "Integrated bioengineering and geotechnical treatments for streambank restoration and stabilization along a landfill." Journal of Soil and Water Conservation. Soil & Water Conservation Society. Vol. 61 No 3, pp. 144-152.

Barrett, Kirk, Krista Reinhart, and Wendi Goldsmith (2003). "Soil Bioengineering Stabilization of an Eroding Streambank along a Sanitary Landfill, Mill Creek, Cincinnati, Ohio." Proceedings, *World Water & Environmental Resources Congress, June 23-26, 2003, Philadelphia, PA*

PROJECT #30: CHARLES RIVER WATERTOWN ARSENAL

PROJECT TYPE: **Riverbank stabilization**

PROJECT SCALE : Medium

CLIENT/OWNER: US Army Corps of Engineers, New England District

TECHNIQUES EMPLOYED: **Coconut Fiber Roll**
Live Brushlayering
Live Staking

GEOGRAPHIC LOCATION: Watertown, MA

Fig. 1 Newly planted wetland shelf with goose fencing

GEOMORPHIC SETTING: The Charles River is located in a watershed with a total area of 308 square miles. The watershed is urbanized, and the river has multiple dams. Beginning in Watertown the river corridor features nearly continuous public parks and greenways, most of which were designed by Frederick Law Olmstead over a century ago. The Charles River is fed by more than 80 streams and tributaries as well as multiple large aquifers. Its mouth at Boston Harbor is controlled by a dam which is used to manage water levels for recreation and periodic flood control, as well as preventing undesirable tidal influence. This slow-moving river exhibits few natural geomorphic processes; it remains nearly fixed in its plan form geometry as the river meanders 80 miles from Hopkington to Boston, connecting 23 communities and recharging many ponds and wetlands before finally discharging its tea colored waters into the sea (Fig. 1).

SITE CONDITIONS AND PROBLEMS: Much of the park land bordering the Charles River was developed based on the principle of "reclaiming" marshlands within its estuarine reaches by placing fill to raise ground surfaces and establishing lawns, trees, and recreational features. The park corridor was viewed as a greenway feature connecting adjacent communities to the river and each other, and creating high property values nearby. The US Army offered the Watertown Arsenal site, including handsome brick buildings on a large acreage abutting the river, to the local community which was eager to acquire and develop it.

W. Goldsmith et al., *Bioengineering Case Studies: Sustainable Stream Bank and Slope Stabilization*,
DOI 10.1007/978-1-4614-7996-3_31, © Springer Science+Business Media New York 2014

The site had been used for munitions testing, storage, and manufacturing for over 150 years, so it came as no surprise that contaminated soils, groundwater, and subsequently surrounding plant life were detected during early inspections. The main contaminants of concern included radioactive isotopes, all of which were fully removed. However, investigations revealed that all of the Olmstead-era fill materials contained low levels of contamination typical of coal ash and burned garbage which had been placed to fill wetlands. Federal regulations dictated full remediation of all contaminants on the federally owned parcel, even though adjacent nonfederal parcels contained identical materials. The site was fenced and access restricted while stakeholders devised a suitable remediation plan (see Fig. 2).

TREATMENT OBJECTIVES AND CONSIDERATIONS: In 1994, this site was placed on the EPA Superfund National Priority List. The remediation challenge was to restore the site's ecological productivity and recreational use while providing for containment of any residual contamination that might still exist. After exploring many options to preserve some areas containing specimen trees and sensitive riparian vegetation, a decision was reached to remove all contaminated soils to a depth of 3 ft, but to replace them with matching terrain and plant species capable of providing habitat and physical protection. Noting the increasing use of power boats on the river, and the widespread undercutting of existing banks, rock was placed at the toe for added armor protection.

Fig. 2 Contaminated fill in Olmstead park site adjoining the Charles River

TREATMENTS SELECTED:

- **<u>Live Brushlayering</u>.** Dormant branches were inserted between layers of earth placed on the bank. This was the main technique used to reestablish vegetation on the bank (see Fig. 3).

- **<u>Erosion Control Blanket</u>.** A coconut fiber erosion control blanket was placed on the upper bank to provide temporary cover and promote the establishment of vegetation.

- **<u>Live staking</u>.** Live stakes were planted through the blanket so new woody vegetation could become established quickly and provide a dominant cover of native woody plants (refer to Fig. 3).

- **<u>Coconut Fiber Rolls</u>.** A layer of rock was placed at the toe to provide scour protection. In turn a line of coconut fiber rolls was placed atop the rock below the normal water line. The role was anchored to function as a breakwater against wave action from passing watercraft. The breakwater protected a littoral shelf where herbaceous plants were added to establish a marsh community.

Fig. 3 Brush layering placed on lower bank below erosion control blankets. Live stakes were driven through the blanket at the top to provide additional protection

OBSTACLES TO IMPLEMENTATION

The wide array of federal oversight agencies, public park stewards, community groups, and NGOs were unfamiliar with the proposed bioengineering treatments and a lot of basic education and outreach were needed. However, in the end highly naturalistic and attractive treatment options allowed the stakeholder group to arrive at consensus on how to proceed. Protective fencing was used to discourage nonmigratory urban geese from uprooting new plantings.

PERFORMANCE EVALUATION

The innovative treatment approach that was adopted resulted in a stabilized layer of soil 3 ft deep along the riverbank which effectively capped and immobilized residual contaminated floodplain materials. Restored habitat areas were protected through the creation of a reestablishment of riparian forest buffer and a herbaceous littoral shelf wetland. Together they collectively helped protect against erosion from waves and runoff (see Fig. 4). The plantings provided rapid reestablishment of plant communities that existed previously within the historic park.

Fig. 4 Recently installed riparian forest buffer and littoral shelf wetland

Fig. 5 Charles River Watertown Arsenal site 12 years after construction

BENEFITS AND LESSONS LEARNED

By cultivating awareness of bioengineering methods on this highly sensitive and regulated segment of the Charles River, similar treatments were adopted on numerous other sites within the park systems. The project became the first Superfund Site in New England to receive bioengineering treatments that functioned to stabilize and cap residual contaminants. At the same time this bioengineering approach also provided mitigation for habitat, water quality, recreation, and historic landscape esthetics. Long-term protection and good vegetative establishment were also achieved at a reasonable cost (see Fig. 5).

ADDITIONAL READING:

USACE (2009). *Report of the Secretary of the Army on Civil Works Activities for FY 2009*. United States Army Corps of Engineers. New England District.

PROJECT #31: CONNECTICUT RIVER

PROJECT TYPE: **Stream bank stabilization**

PROJECT SCALE: **Large**

CLIENT/OWNER: Western Mass. Electric Company

TECHNIQUES EMPLOYED: **Live Brushlayering**
Live staking
Vegetated MSE

GEOGRAPHIC LOCATION: Northfield, MA

Fig. 1 View of Connecticut River

GEOMORPHIC SETTING: At the project location, the Connecticut River has an annual average discharge of 14,128 cfs, a drainage area of 7,181 square miles. The river banks are steep and unstable; they are composed of non-cohesive soils derived mainly of glacial outwash, deltas, and ancient postglacial dune formations. The project reach is influenced by the presence of hydropower dams upstream and downstream. A pumped storage facility used for managing peak power generation creates additional patterns of water level change. Two significant factors causing bank erosion are weekly fluctuation in water level of 3–5 ft caused by complex hydropower operations and boat wakes of 0.5–1.3 ft from recreational traffic (Fig. 1).

SITE CONDITIONS AND PROBLEMS: Scour and piping at the bank toe was exacerbated by constant water level fluctuations. These in turn undercut the toe causing upper banks to exhibit slow raveling of their surfaces or abrupt block failure, both of which resulted in the loss of mature stabilizing trees on the banks.

Prior attempts to control erosion included a range of unsatisfactory practices such as helicopter logging of the banks based on the idea that tree weight caused bank collapse, armoring banks with cable-tied used tires which were later considered illegally disposed waste. In addition riprap was placed on extensive areas, much of which sank into the deep alluvial sands along the steep banks. These prior methods were expensive and had either failed to function effectively or were ecologically or esthetically unacceptable. Without addressing these erosion problems, hydropower facilities would not achieve relicensing.

DOI 10.1007/978-1-4614-7996-3_32,

TREATMENT OBJECTIVES AND CONSIDERATIONS: The Connecticut River has been the focus of erosion control efforts for two decades as use of the river for recreation and hydropower has increased, and stakeholder concerns were addressed. The project required enhancement of habitat conditions, compatibility with recreational uses, and avoidance of flood storage displacement, while providing over 5,000 linear feet of bank stability and protection. Preservation of the existing narrow wooded buffer, adjacent agricultural lands, and Native American artifacts were a priority. The unique hydrologic regime required careful selection of vegetation tolerant of water level fluctuations. Designs focused on preventing erosion and undermining at the toe of the steep banks.

TREATMENTS SELECTED:

- **Mechanically Stabilized Earth (MSE).** Various reinforcing materials were used to build stable bank profiles where soil fills were added to the bank. In lower bank zones susceptible to periodic flood inundation, geogrid was used to wrap lifts of soil. Higher on the bank where inundation was infrequent and tractive stresses were lighter, coconut fiber mesh was used to wrap and stabilize lifts of soil. At one site where a permanent slope face exceeding 1:1 was required to protect a state champion tree, a twisted wire mesh material was used to provide mechanical reinforcement at the face and between soil lifts.

- **Live Brushlayering.** By selecting plants that were native to the project area, a suitable palette of species was employed. In addition to typical materials such as willow and dogwood, sycamore, poplar, box elder, and silver maple were also used. In most cases these were combined with MSE treatments.

- **Live Staking.** Dormant cuttings were typically planted into banks that had already been stabilized by other measures. These cuttings were installed during the dormant season (see Fig. 2). Live stakes were primarily derived from willow, alder, dogwood, and elderberry.

OBSTACLES TO IMPLEMENTATION

The most severe challenge to accomplishing construction was limited access to project sites. Top of bank conditions were both precarious and hence hazardous for heavy equipment; furthermore, in most cases sensitive resources were located there that needed to be avoided. Sediments below steep bank areas were soft and exhibited low bearing capacity. River levels fluctuated often, so access by barge

Fig. 2 Live stakes installed planted on MSE treated bank

from the water side was unreliable, though possible at times, mainly for placement of stone toe protection. In nearly all cases, bank reconstruction required creation of soil benches which served as temporary access and staging areas. Property owners each had their own, often vastly different, preferences and restrictions regarding access, timing, and treatment type. The scale of the project was large enough, and spanned multiple sites, so that portions of the work were planned for nondormant seasons. On the other hand, the installation and planting of bioengineering materials (transplants and live cuttings) was carried out during the proper dormant season.

PERFORMANCE EVALUATION

Since its inception, the project was subjected to extensive assessment by members of the Connecticut River Erosion Committee established as a community oversight body by the Federal Energy Regulatory Commission. The project received formal post-construction monitoring that was funded through Massachusetts DEP and US EPA. Survival of vegetation was in the 50–90 % range, which was considered appropriate based on species and setting. The bank was stable and protected by a dense cover of vegetation which was well adapted to fluctuations in river water level (see Fig. 3). Reversal of the prior pattern of bank retreat, sedimentation is evident on and in front of virtually all banks (see Fig. 4).

Fig. 3 A robust cover of diverse vegetation after 2 years

Fig. 4 After one full year, plants flourish and toe is stable and accreting sediment

BENEFITS AND LESSONS LEARNED

In addition to delivering cost-effective stabilization treatments, the project has been used to provide demonstrations of various approaches with site-tailored habitat functional value. Future phases of the project have been implemented, and similar measures have been adopted for other large rivers throughout the northeast.

<u>ADDITIONAL READING</u>:

Goldsmith, Wendi (1998). "Soil Reinforcement by River Plants: Progress and Results." *Engineering Approaches to Ecosystem Restoration, Conference on Wetlands Engineering & River Restoration : March 22-27, 1998, Denver, Colorado*.

PROJECT #32: CUMBERLAND RIVER SHELBY BOTTOMS PARK

PROJECT TYPE: **Bank stabilization**

PROJECT SCALE: **Large**

CLIENT/OWNER: U.S. Army Corps of Engineers, Nashville District

TECHNIOUES EMPL0YED: **Willow Poles, Live Staking, Brush Layering, Live Fascines ECBs**

GEOGRAPHIC LOCATION: **Nashville, TN**

Fig. 1 View of Cumberland River

GEOMORPHIC SETTING: The Cumberland River stretches for almost 700 miles through Kentucky and Tennessee including lands with some of the greatest biodiversity in the USA. The river exhibits a highly sinuous pattern, with steep banks and little effective floodplain due to inundation from a nearly continuous series of large dams built primarily for navigation purposes. The river itself has been subjected to many modifications, as well as urban development and ensuing suburban sprawl throughout much of the watershed. Heavy barge traffic is common on the highly managed river that features locks and dams. Soils mainly consist of heavily weathered clays.

SITE CONDITIONS AND PROBLEMS: A local farm was converted to a major park on the outskirts of Nashville, with the intent of providing access to passive recreation, natural trails, watershed education, and demonstration of green infrastructure measures. In many locations, park trails and existing or planned structures, such as observation decks, were jeopardized by severe erosion of the banks of the Cumberland River due to heavy wake from large barges (Fig. 1). Dozens of trees had fallen into the river, experienced severe root scour, or were undermined and threatened by bank erosion.

W. Goldsmith et al., *Bioengineering Case Studies: Sustainable Stream Bank and Slope Stabilization*,
DOI 10.1007/978-1-4614-7996-3_33, © Springer Science+Business Media New York 2014

TREATMENT OBJECTIVES AND CONSIDERATIONS: The project area involved the design of measures for the restoration and protection of a 3,000-ft reach of the Cumberland River along Shelby Bottoms Park. A site assessment was conducted followed by hydraulic analysis that included modeling the effects of heavy barge traffic on the banks in order to establish the basic remediation design criteria. Three different bioengineering treatments were prepared, employing techniques such as Willow Poles inserted through a rock toe and gabion mattress, Live Staking, Live Brushlayering, Live Fascines, and Erosion Control Blankets. Each were configured to match bank conditions and to allow viewing corridors from elevated platforms used by park visitors. A heavy rock toe was placed to provide stability for the low bank area where plant roots could not effectively stabilize soil and where wave impacts from barges created intense erosive forces.

TREATMENTS SELECTED:

- **Willow Poles and Live Staking.** Willow and cottonwood trees may be propagated by planting dormant cuttings. These reinforce and strengthen the soil as roots develop and are secure and stable when installed. Although smaller (<4 cm (1.5 in.)) diameter cuttings (stakes) grow more vigorously than older, larger materials (poles), larger materials provide better mechanical bank protection during the period of plant establishment. Dense arrays of poles reduce local near-bank water velocities.

- **Live Fascines.** The design called for extensive use of willow and button-bush, which were prevalent and easily harvestable in the area. An Erosion Control Blanket was placed on the slope surface and also run into shallow trenches dug on contour. Live Fascines were placed in the trenches and staked in place to further secure the blankets (see Fig. 2). The Live Fascines firmly anchored the blankets to the slope and prevented rills or small gullies from developing.

- **Live Brushlayering.** On the steepest and tallest slopes which included evidence of internal seepage, a more complex bank treatment was selected. The slope was benched and compacted, and live brush was added in layers prior to placing additional fill to create final grade (see Fig. 3).

Fig. 2 Live Fascines run in trenches and staked in place

Fig. 3 Live Brushlayering placed diagonally on benched slope

OBSTACLES TO IMPLEMENTATION

Bank stabilization at this site was the second major application of bioengineering treatments at an Army Corps of Engineers project completed by the Nashville District. The materials and methods remained unfamiliar to many of the parties involved with contracting, permitting, and construction. Access to the bank was limited in order to preserve trees remaining on the top of bank in most areas, and work was completed partly from barge (rock placement for toe protection) and partly using manual labor on a steep bank above the stone toe and below existing large riparian trees.

PERFORMANCE EVALUATION

The various bioengineering treatments established rapidly and performed successfully. Ongoing pruning of dense brush and tree cover is required to maintain views in selected areas. Shrub species, in particular buttonbush, has become established in many portions of the stone toe, adding a level of biological productivity and esthetic value that exceeded the original expectations.

BENEFITS AND LESSONS LEARNED

One innovative practice the contractor employed was the use of boards, held in place temporarily with rebar, to shape and retain soil benches during the compaction process.

PROJECT #33: MANHAN RIVER

PROJECT TYPE: **Bank stabilization**

PROJECT SCALE: **Medium**

CLIENT/OWNER: Tennessee Gas Pipeline Company

TECHNIQUES EMPLOYED: **Coconut Fiber Rolls, Bendway Weirs Rock Vanes**

GEOGRAPHIC LOCATION: Easthampton, MA

Fig. 1 Coconut fiber roll

GEOMORPHIC SETTING: The Manhan River is a 28-mile-long river in western Massachusetts and a tributary of the Connecticut River. Most of its watershed contains forest cover, with farmland and low density suburban development in some areas. These features make it a healthy and resilient river with high biodiversity. Most of the channel exhibits stable meandering alluvial geomorphic patterns (Fig. 1).

SITE CONDITIONS AND PROBLEMS: The river had become unstable where a natural gas pipeline was threatened with exposure by a poorly vegetated rapidly eroding outside bend. Past construction and farming in this area had removed stabilizing forest cover. These conditions triggered an emergency which needed an immediate solution. Because of the presence of five endangered species, regulators required an ecologically sensitive approach and denied an initial proposal for rock armor.

W. Goldsmith et al., *Bioengineering Case Studies: Sustainable Stream Bank and Slope Stabilization*,
DOI 10.1007/978-1-4614-7996-3_34, © Springer Science+Business Media New York 2014

TREATMENTS SELECTED:

- **Coconut Fiber Rolls.** The bank reconstruction included placement of Coconut Fiber Rolls in order to stabilize the bank without the presence of trees within the right-of-way for the high pressure gas pipeline. This treatment protects the banks by providing short-term soft armor and long-term reinforcement of grass root structures, but, unlike riprap, does not lead to increased downstream erosion because of the roughness it adds.

- **Erosion Control Blankets.** An erosion control blanket was placed above the fiber rolls to allow bank reshaping and establishment of warm and cool season grass seed.

- **Rock Vanes.** These stone structures placed on the outside bend to help dissipate energy and deflect water flow away from the bank. These actions decreased tractive stresses near the bank toe and promoted the capture of sediment so the rate of erosion near the pipeline was significantly reduced (see Figs. 2 and 3).

Fig. 2 Rock vane pointing upstream redirects flow away from bank

Fig. 3 A series of Rock Vanes located on an outside bendand pointing upstream redirect flow away from bank

OBSTACLES TO IMPLEMENTATION: Because the risk to the pipeline was discovered in autumn, the project had to be completed rapidly in order to avoid shutting off gas service to the region during peak heating season. State regulators were concerned that hasty and poorly selected and executed treatments would lead to short- and long-term impacts to the river and associated habitat. The design and construction team had to demonstrate its credibility in order to negotiate emergency permits and persevere despite an earlier cease and desist order. Care was also required not to damage the habitat of five endangered species.

PERFORMANCE EVALUATION

The channel is now stable; no risk to the pipeline has been identified since completion. The banks are well vegetated and no longer eroding (see Fig. 4). The private property owner is also satisfied with the outcome of the work and the function and appearance of the river.

Fig. 4 Vegetated and well-protected stream bank after repair and stabilization

BENEFITS AND LESSONS LEARNED

The avoidance of in-stream excavation, the increase in plant cover, and the incorporation of bendway weirs that could be used as resting perches enhanced habitat conditions for endangered fish, dragonfly, and mussel species of concern. The gas pipeline company became acquainted with an ecologically sensitive approach to addressing even the most severe threats to pipeline safety and learned that bioengineering measures could receive faster regulatory approval than hard structures.

PROJECT #34: WALGREENS SLOPE

PROJECT TYPE: Slope stabilization

PROJECT SCALE: Medium

CLIENT/OWNER: BRT Management

TECHNIQUES EMPLOYED: **Live Fascines**, **Live Pole Drains**, **Live Staking**, **ECBs**

GEOGRAPHIC LOCATION: Marshfield, MA

Fig. 1 Slope failures and erosion

GEOMORPHIC SETTING: The site is part of the landform creating Cape Cod. Area soils consist of glacial outwash and moraine deposits, many of which form steep noncohesive slopes. Area soils permit infiltration of even the highest rainfall volumes when they are covered with forest vegetation and leaf litter. However, if the forest cover is disturbed, many local soil types become highly susceptible to both surficial erosion and to deeper seated slope failures (Fig. 1).

SITE CONDITIONS AND PROBLEMS: In the process of preparing a site for the development of a retail plaza, vegetation and soils were disturbed, triggering substantial erosion. After one rainstorm, widespread rill and gully formation, as well as slumping of large blocks of soil, resulted in the building site being covered with sediment. Efforts to stabilize the 30 ft high slope with hydroseeding and other superficial treatments failed repeatedly. The civil engineering consultants had identified retaining costly wall systems as an alternative; however, local regulatory bodies would not permit this alternative.

TREATMENT OBJECTIVES AND CONSIDERATIONS: Regulatory agencies required the developer to employ an ecologically functional approach to stabilizing the slope with vegetation as opposed to relying on a hard structure. The goal was to

DOI 10.1007/978-1-4614-7996-3_35, © Springer Science+Business Media New York 2014

maintain site esthetics and provide linkage to habitats including a large adjacent oak and pine-dominated woodland. All plantings had to be adapted to the sandy and acidic soil present, which was typical of the area. Runoff from heavily used trail systems uphill from the impacted slope had to be managed, and temporary stabilization measures were needed to match the physical challenges presented by steepness of the slope. Additionally, native plant species were mandated so as to be consistent with surrounding natural forests and the general visual character of the surrounding urban area.

TREATMENTS SELECTED:

- **Live Pole Drains.** Live fascines were secured in place in trenches to intercept surface and subsurface water and carry it downslope in a controlled manner (see Fig. 2). Soil backfill is retained by vegetation, which requires soil contact to survive and sprout. The system is integrated with erosion control blankets. Willow and alder materials function initially as preferential flowpaths for water, and over time as roots and shoots develop, the trench maintains or improves its stability through development of roots and shoots.

- **Live Fascines.** Fascines of willow and alder were secured on contour at various locations along the slope to break the slope into shorter segments, where runoff could be intercepted and allowed to infiltrate into the ground.

Fig. 2 Facines installed as Live Pole Drain for capturing and conveying water downslope

- **Erosion Control Blanket.** The entire regraded slope was prepared with a quality topsoil, seeded with warm and cool season grasses, and covered with a durable erosion control blanket. The blanket was fabricated from coconut fiber and secured in place on the slope with biodegradable stakes. Care was taken to fit blankets firmly to soil surface, with adequate overlap and attention to detail at key junctures such as the top and sides of slope, or where fascines and other measures intersected one another.

- **Live Staking.** Regrading and blanket placement were carried out during the summer to provide short-term protection and emergency slope repair work. Native species including willow, alder, and viburnum cuttings were added through dormant stake placement later in the fall (see Fig. 3).

Fig. 3 Site with key materials in place, awaiting dormant period for fascine and stake placement

OBSTACLES TO IMPLEMENTATION

The project owner had a limited construction budget, much of which had already been expended on futile attempts. The key parties responsible for design and over-sight of construction, as well as compliance with regulatory requirements, were unfamiliar with bioengineering measures. These parties initially did not understand how such seemingly simple measures could outperform earlier efforts. Extensive communication, persuasion, and documentation were needed in order to promote acceptance of bioengineering solutions as a viable alternative to a structural retaining wall.

PERFORMANCE EVALUATION

Bioengineering treatments were executed on an emergency basis, outside the standard dormant period, using phasing to allow dormant woody plants to be added subsequently. Nevertheless, no surface or deep seated slope failure occurred after treatments were initially placed. The site was rapidly stabilized and protected by a vigorous vegetative cover (see Fig. 4).

Fig. 4 Site after first winter shows stability and growth

BENEFITS AND LESSONS LEARNED

From the site owner's point of view, the project was a success due to compliance with regulatory requirements and stabilization of the eroding slope, which allowed building construction and overall site development to proceed. Conservation and community planning interests were satisfied that native plants and a visually compatible design were used. All parties were interested to learn how a cost-effective treatment could be achieved through proper planning and execution of soundly designed bioengineering methods. Local regulators expressed an interest in clarifying and adopting higher standards for similar projects in the future.

PROJECT #35: CREEK ROAD

PROJECT TYPE: **Stream restoration and bank protection**

PROJECT SIZE: Medium

CLIENT/OWNER: Town of Yorkshire

MEASURES: **Rock vanes, Bendway weirs, LSTP, Revegetation**

LOCATION: Cattaraugus Creek at Creek Road Yorkshire, NY

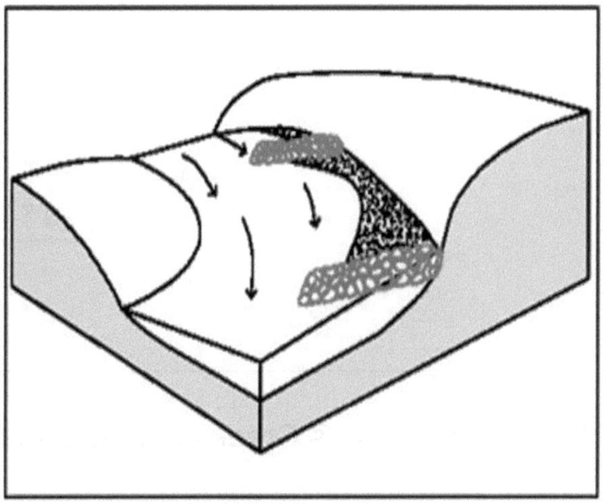

Fig. 1 Schematic of bendway weir

GEOMORPHIC SETTING: Creek Road parallels Cattaraugus Creek, which meanders through its valley as it flows from east to west. Creek Road is situated on left valley side and is within 30 ft of Cattaraugus Creek in the vicinity of the project. The watershed upstream of the project is approximately 187 square miles. Cattaraugus Creek at Creek Road has a bankfull flow (i.e., Q2, or 2-year recurrence interval flow) of approximately 12,400 cfs, and the Q100 is approximately 19,100 cfs. The geologic setting of the stream varies over its length, but it generally flows through alluvial flood plain deposits containing varying amounts of silt, sand, and gravel. The underlying strata are also comprised of glacial outwash deposits of sand and gravel and glacial tills that are dense deposits of clay, silt, sand, and gravel, in addition to occasional bedrock outcrops.

SITE CONDITIONS AND PROBLEMS: Creek Road is located at the top of a bank at the outside bend of one of the stream's meanders. The stream has changed course frequently over time (refer to Fig. 2) and has encroached on adjacent infrastructure. Previous efforts to protect and support the highway embankment using a tied-back sheetpile wall bulk-head were unsuccessful (refer to Fig. 3). Water flow eroded toe support in front of the wall and approximately 280 ft of the 570-ft long wall eventually collapsed. Erosion of the bank continued as the stream meandered toward Creek Road. Streamflows of approximately 16,000 cfs from a storm event with a recurrence interval of 45 years in August of 2009 caused erosion of the bank, encroaching to within a few feet of the road's shoulder.

W. Goldsmith et al., *Bioengineering Case Studies: Sustainable Stream Bank and Slope Stabilization*,
DOI 10.1007/978-1-4614-7996-3_36, © Springer Science+Business Media New York 2014

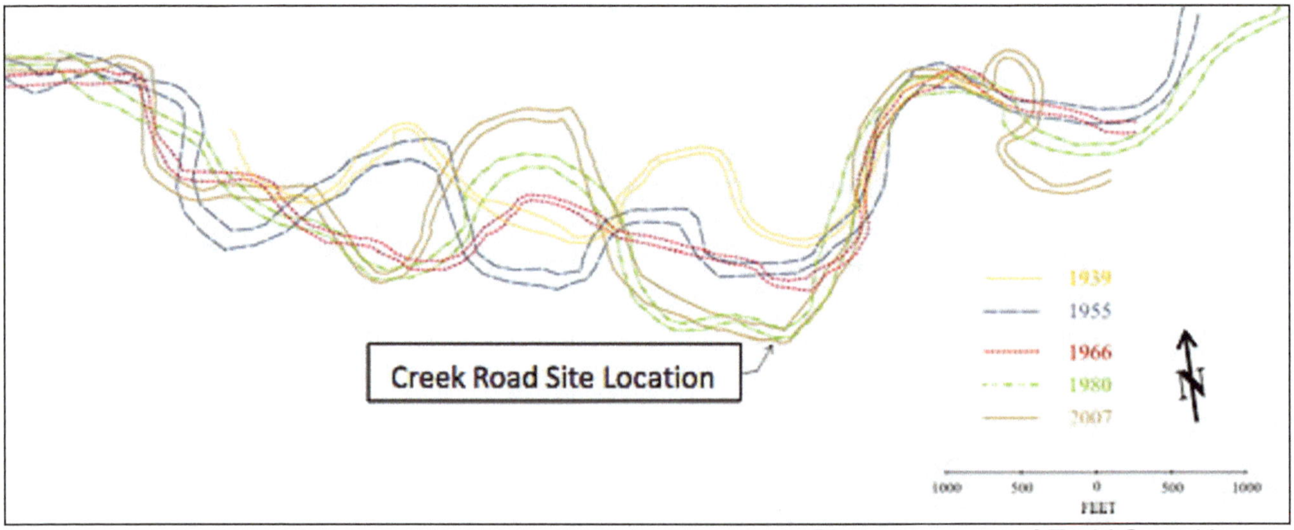

Photo courtesy of Todd Swackhamer

Fig. 2 Cattaraugus Creek channel planform adjustment in the reach near the project site

Photo courtesy of Todd Swackhamer

Fig. 3 Looking upstream at the failed sheetpile wall prior to remediation

The bank at the project site is underlain by sandy alluvium soil deposits with varying amounts of gravel, silt, and clay. Occasional layers of silt and clay lake deposits were encountered during the subsurface exploration program. Generally, the plastic deposits are stiff, and the granular deposits are loose to medium dense. In general groundwater elevations in the bank are approximately 1 ft above the adjacent stream level. Slope stability analyses indicated that the banks had factors of safety against deep and rotational failures greater than unity. However, analyses of surficial failures following removal of support from the bank's toe had factors of safety approaching 1.0. These results indicate that the subsurface conditions are generally sufficient to support the stream bank, except following high flow conditions that eroded the bank's toe.

TREATMENT OBJECTIVES AND CONSIDERATIONS: The main objective was to protect the left bank, between Cattagaugus Creek and Creek Road, from further erosion and to stabilize the road's subgrade. Since part of the solution involved flattening the bank slope, the toe of the bank at the water line was moved further out into the channel. This required some of the construction work to be completed in the stream. As such, the designers wanted to create a solution that was sensitive to the flora and fauna in and around the stream.

Furthermore, the designers wanted to avoid displacing the problem at this site to another location immediately upstream or downstream of the project. Therefore, the solution had not only to resist the erosive forces of the stream but also reduce some of flowing water's energy and redirect it away from other property vulnerable to damage.

To accomplish this, the project was designed to use the best construction practices for bank protection, as found in the Federal Highway Administration (FHWA), Hydraulic Engineering Circular (HEC) series of publications, as well as various other professionally accepted documents (Arneson et.al. 2012; Biedenharn et.al. 1997; Derrick and Freeman 2004; Lagasse et al. 2001; McCullah and Gray 2005; Rosgen 1996).

TREATMENTS SELECTED:

The structural and biological components of the stream bank protection system consisted of a stone fill false bankline, improved drainage, a flattened bank slope, willow and dogwood pole plantings, reuse of the remaining sheetpile wall for toe protection, redirective structures (vanes and bendway weirs), and an upstream flank key. Refer to Fig. 4 for the remediation plan.

False bankline (LSTP). A false bankline was created to realign the left bank. The bank was aligned as closely as practicable to the proposed thalweg to convey water smoothly around the bend. The bankline was constructed of stone fill sized for the anticipated streamflows. This provided a longitudinal stone toe protection (LSTP) of the bank along the project limits. The cross-sectional area of the stone fill was large enough to account for "launchable" riprap along the toe as the stream bed geometry changes over time.

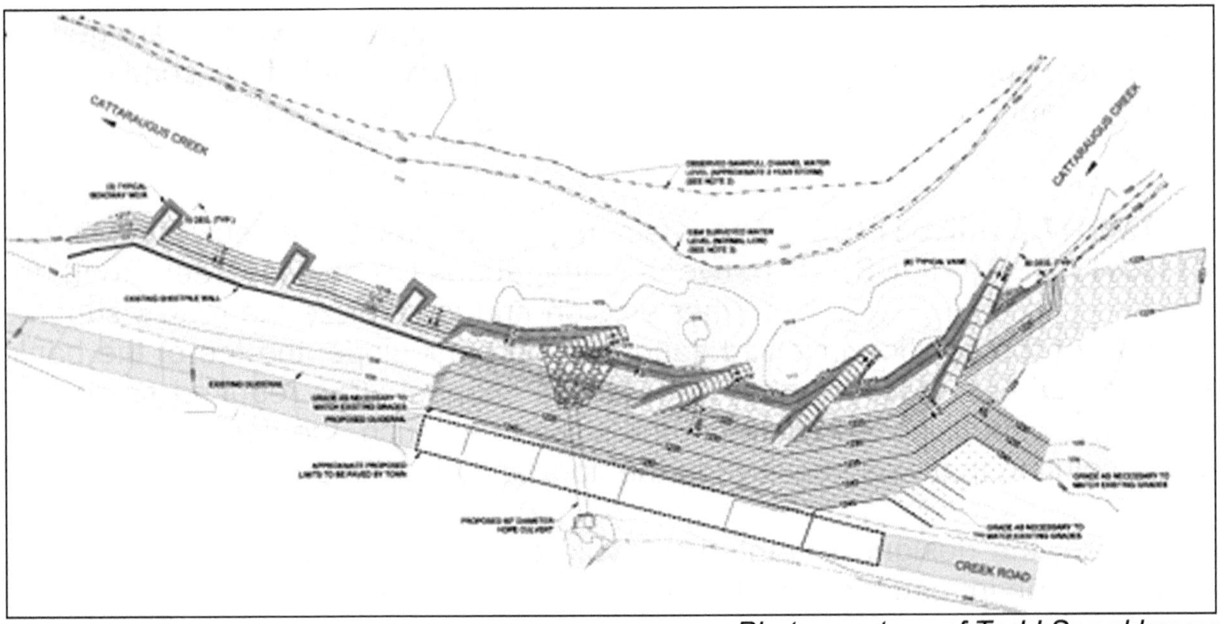

Photo courtesy of Todd Swackhamer

Fig. 4 Remediation plan (final conditions)

Slope drainage measures and revegetation. The area behind the false bankline, and in front of the eroded bank, was backfilled with compacted sand and gravel structural fill. Drainage improvements, such as trench drains and an open stone blanket as well as pole drains, were also included to transmit groundwater out of the bank. The false bankline provided sufficient area to flatten the bank to approximately 2.5(H):1.0(V) or less. The flattened slope was vegetated with root mass grass seed and willow and dogwood cuttings above the Q10 flow elevation.

Redirective structures. Rock vanes and bendway weirs were used as redirective measures. The vanes and weirs are stone fill "piers" or "fingers, which extend from the bankline into the channel in an upstream direction. They are designed to redirect the thalweg and reduce the flow energy near the bankline. The project used four rock vanes along the upstream portion of the project, at the beginning of the stream bend in a scour hole. The vanes are oriented at 30° from the bankline, spaced approximately 70–80 ft apart, and are 40–60 ft long. These vanes were designed to direct the thalweg's alignment away from the bankline as it flows around the bend. Three bendway weirs (refer to Fig. 1) redirect the thalweg away from the bank at the downstream end of the project. The weirs are oriented at 70° from the bankline, spaced approximately 80 ft apart, and are 20 ft long. They were designed to maintain the thalweg's alignment near the center of the channel as it exits the project limits.

Upstream flank key. The upstream flank key is a trench that extends (in plan) from the upstream end of the project, inland at an approximately a 30° angle from the bank. The flank key is intended to protect the project from being bypassed from behind the erosion protection. The upstream flank key was excavated to an elevation 2-ft below the bottom of the stream and backfilled with stone fill.

OBSTACLES TO IMPLEMENTATION: A diagonal bar of sediment accumulated on the bed of the channel immediately upstream of the site. The bar spanned the width of the channel and directed the flow toward the left bank, creating a scour hole on left-approaching bed. A topographic survey of the site confirmed that, under an annual-low, base-flow water surface, a 5-ft deep scour hole was present. This scour hole had to first be filled with stone fill to be able to create a stable false bankline.

A steep, valley-side tributary discharges through a 60-in. culvert onto the left bank, midway through the bend. The culvert discharges onto the upstream slope of the downstream-most vane, where it deposits its bedload sediment. This tributary was actively eroding through glacial outwash and till, and it delivered the majority of the sediment that had accumulated behind the sheetpile wall, thus contributing to its rotational failure. The design had to incorporate additional erosion protection and sediment control at the culvert's outlet.

PERFORMANCE EVALUATION: A storm hydrograph for Cauttaraugus Creek since completion of construction (see Fig. 5) was reviewed. The gaging station for the hydrograph is in Gowanda, New York, approximately 31 miles downstream of the site. It is the only operable gaging station on Cattaraugus Creek. The hydrograph shows that Cattaraugus Creek has exceeded bankfull flows several times since the construction at Creek Road was completed, and that there have been numerous storm flows in the basin. Observations made during and after the bankfull flows indicate that the redirective field is functioning as intended.

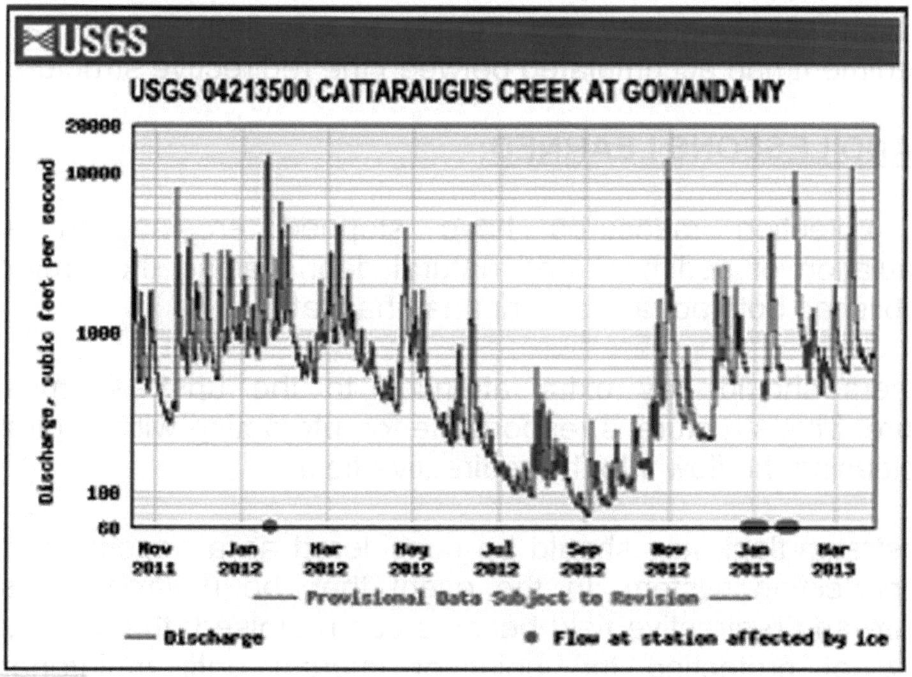

Fig. 5 Flow in Cattaraugus Creek since construction completed (2011)

Photo courtesy of Todd Swackhamer

Fig. 6 Looking downstream following remediation during normal flow conditions

The upstream-most vane intercepts the force of the streamflow and releases it into the redirective field of the remaining downstream vanes and weirs. The second vane lies in the hydraulic shadow of the first vane, and the discharge flows neatly past the second and over the third vanes. The fourth vane receives the flow from the third and directs the flow into the field of bendway weirs. The bendway weirs force the thalweg down the center of the channel on exiting the bend. Observations at the site show these redirective structures are working as designed during both normal and high flow conditions (see Fig. 6). In less than 2 years, an appreciable amount of sediment had accumulated between the redirective structures.

BENEFITS AND LESSONS LEARNED:

1. Bank protection measures have stopped erosion. In addition, the accumulation of sediment and natural debris along the project reach has reestablished both aquatic and riparian habitat.

2. Designers should pay extra attention to the upstream-most redirective structure. This structure is responsible for interacting with natural flow regime and releasing the flow into the redirective field.

3. An upstream-flank key should be considered as a necessary first step in a bank protection system. In the event that the upstream-most redirective structure and redirective field become compromised, the flank key should be capable of protecting the point of interest until maintenance can be performed.

ACKNOWLEDGMENT

The project description and write-up were provided by Todd Swackhamer, Proj. Manager, and James Ramona, Proj. Geologist, McMahon and Mann Consulting Engineers. Their contribution and assistance are greatly appreciated and acknowledged.

ADDITIONAL READING:

Arneson, L.A., Zevenbergen, L.W., Lagasse, P.F., and Clopper, P.E. (2012). Hydraulic Engineering Circular No. 18 (HEC-18) – *Evaluating Scour at Bridges – Fifth Edition*, U.S. Department of Transportation – Federal Hwy Administration

Biedenharn, D.S., Elliot, C.M., Watson, C.C. (1997). *The WES Stream Investigation and Streambank Stabilization Handbook*. U.S. Army Corps of Engineers, Waterways Experiment Station, Vicksburg, Mississippi

Derrick, D.L., Freeman, G.E. (2004). *Stream Investigation, Stabilization, and Restoration*. Seminar offered by the American Society of Civil Engineers. River Research and Design, Inc.

Lagasse, P.F., Schall, J.D., and Richardson, E.V. (2001). *Hydraulic Engineering Circular No. 20 (HEC-20) – Stream Stability at Highway Structures – Third Edition*, U.S. Department of Transportation – Federal Hwy Administration

McCullah, J., and Gray, D.H. (2005). *Environmentally Sensitive Channel- and Bank Protection Measures*. National Cooperative Highway Research Program Report 544. Transportation Research Board.

Rosgen, D. (1996). *Applied River Morphology*. Wildland Hydrology.

APPENDIX 1

SUMMARY OF ENVIRONMENTALLY SENSITIVE HILLSIDE AND STREAM BANK PROTECTION MEASURES
(adapted from NCHRP, 2005)

RIVER TRAINING STRUCTURES

BENDWAY WEIRS 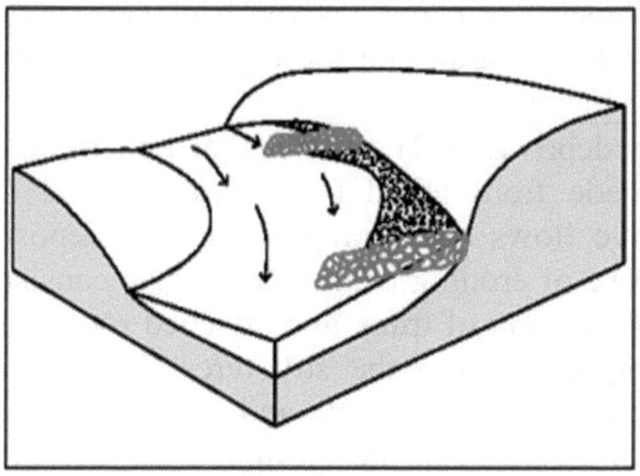	An array of bendway weirs captures and redirects flow at the upper end of a bend, controls the flow throughout the bend, and aims the flow in a direction perpendicular to the last weir in the bend. A minimum of five structures are typically placed in series (the series are known as "weir fields") along straight or convex banklines where flow lines are roughly parallel to the bank. Bendway weirs are generally longer (1/3–1/2 stream width) and lower than barbs or spurs, and are designed to be continuously submerged or at least overtopped often. Bendway weirs also differ from spurs in that they form a control system that captures and directs the stream flow through the weir field, usually all the way through the bend (hence the name bendway weirs).
SPUR DIKES 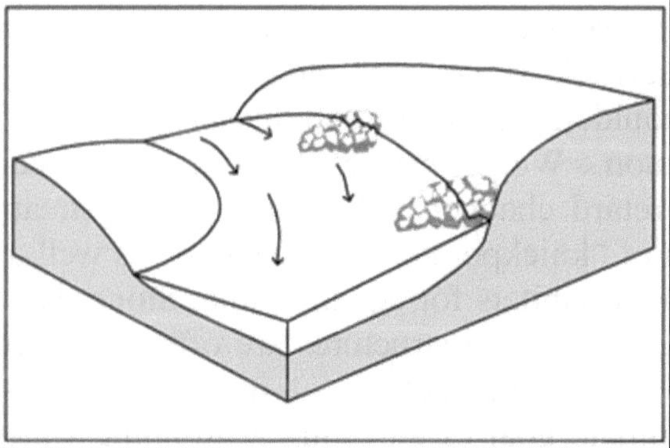	Spur dikes are groins that extend into the stream from the bank to prevent erosion by forcing flow away from the bank. From two to five structures are typically placed in series along straight or convex banklines where flow lines are roughly parallel to the bank. Earthen core spur dikes are groins constructed with a soil core armored by a layer of stone and an embedded stone toe on the upstream side. Stone spurs capped with a prism of earth reinforced with live fascines are referred to as "live booms." Spurs dikes can also be constructed from natural materials, such as large woody debris (LWD), and designed to provide biologic benefits and habitat restoration.
LONGITUD STONE TOE W/ SPURS 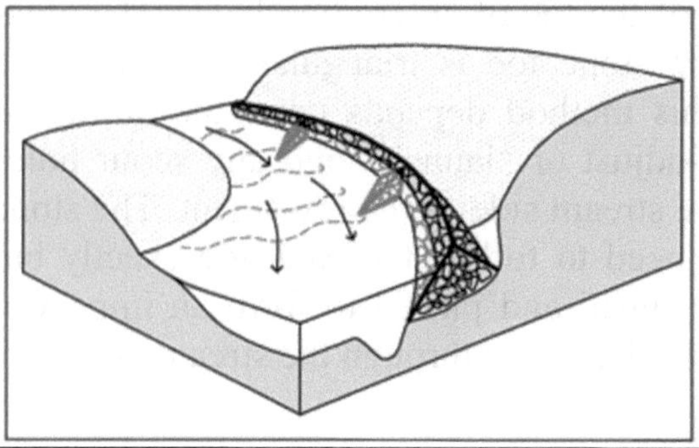	A longitudinal stone toe has proven cost-effective in protecting lower banks and creating conditions leading to stabilization and revegetation of steep, caving banks. A large body of evidence indicates, however, that inter-mittent structures such as spurs tend to provide aquatic habitats superior to those adjacent to continuous structures like a stone toe. This technique represents an effort to achieve erosion control benefits available from a continuous stone toe and habitat benefits associated with spurs.

W. Goldsmith et al., *Bioengineering Case Studies: Sustainable Stream Bank and Slope Stabilization*,
DOI 10.1007/978-1-4614-7996-3,

VANES 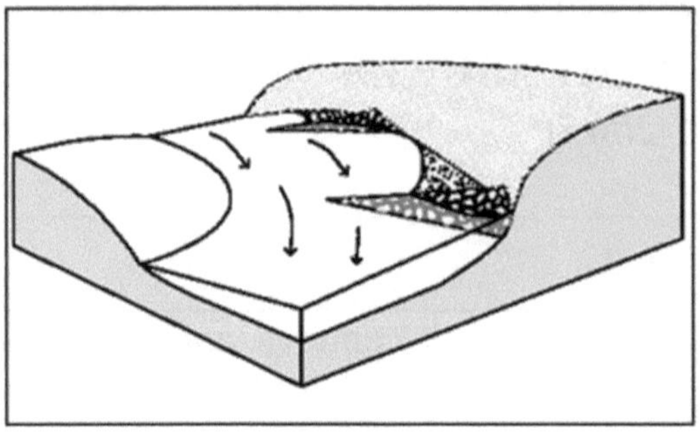	Rock vanes are transverse structures angled into the flow in order to reduce local bank erosion by redirecting flow from the near bank to the center of the channel. River ward tips of the structures are typically low enough to be overtopped by most flows and crests slope upward to reach bank-full stage elevation at the key. Vanes are installed to provide toe protection and rectify lateral instability by redirecting flow away from eroding banks, while providing greater environmental benefits than a stone blanket or revetment. Vanes can increase cover, backwater area, edge or shoreline length, and the diversity of depth, velocity, and substrate.
LWD STRUCTURES 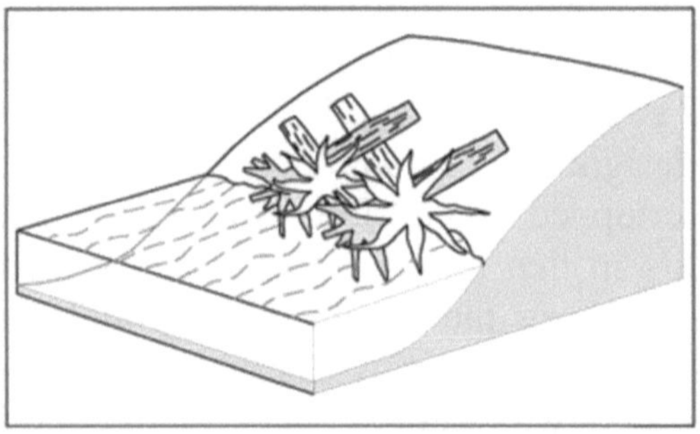	Large woody debris (LWD) structures (aka engineered log jams) made from felled trees may be used to deflect erosive flows and promote sediment deposition at the base of eroding banks. Root wads consisting of a short section of trunk and attached root bole can be incorporated into the structures. Using the classical spur design criteria and methods, the placement of LWD can be designed to achieve optimum benefit for both aquatic habitat and bank protection.
STONE WEIRS 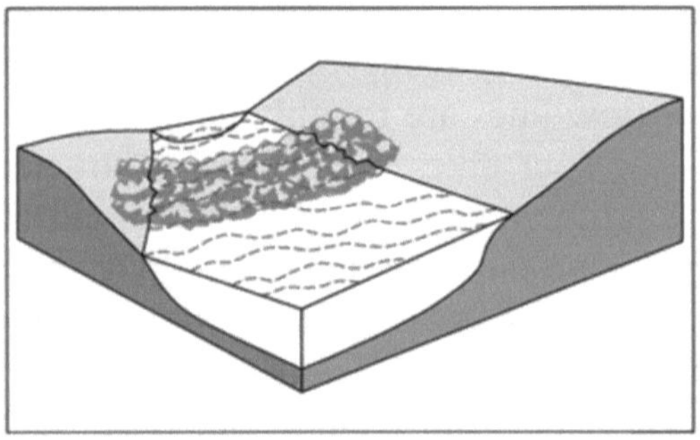	The term "weir" refers to any structure spanning the stream that produces a drop in the water surface elevation. These structures are frequently made of angular quarried stone, but logs, sheet piling, concrete, boulders, gabions, and masonry are also quite common. Well-constructed stone weirs can prevent or retard channel bed erosion and upstream progression of "knickpoints" and headcuts, as well as providing pool habitats for aquatic biota. Stone weirs or similar grade control structures are often intended to raise or elevate the bottom of incised channels, with the ultimate goal of elevating a dropping water table.
LONGITUDINAL STONE TOE 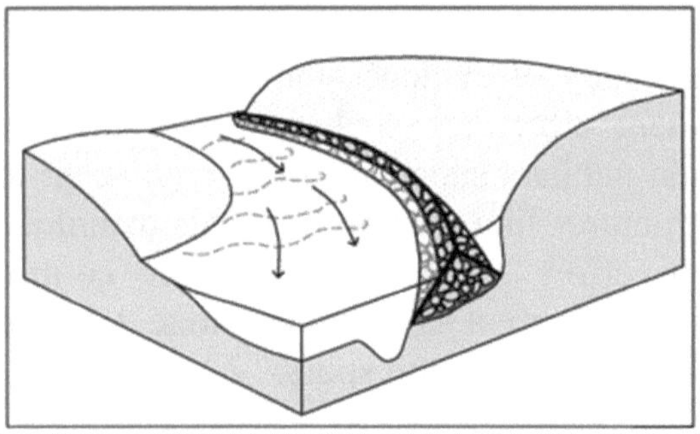	A stone toe is continuous bank protection consisting of a stone dike placed longitudinally at or slightly stream-ward of the toe of an eroding bank. The cross section of the stone toe is triangular in shape. The success of this method depends upon the ability of stone to self-adjust or "launch" into any scour holes formed on the stream side of the revetment. The stone toe does not need to follow the bank toe exactly but should be designed and placed to form an improved or "smoothed" alignment through the stream bend.

VEGETATED MSE 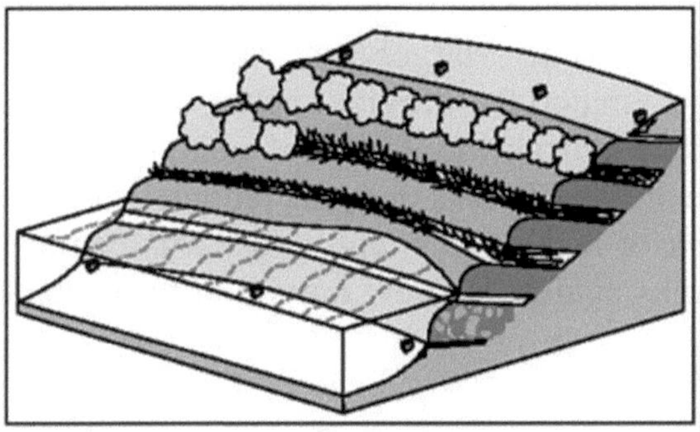	This technique consists of live cut branches (brush-layers) interspersed between lifts of soil wrapped in natural fabric, e.g., coir or synthetic geotextiles or geogrids. The live brush is placed in a criss-cross or overlapping pattern atop each wrapped soil lift in a manner similar to conventional brushlayering (see Technique: Live Brushlayering). The fabric wrapping provides the primary reinforcement in a manner similar to that of conventional mechanically stabilized earth (MSE). The live, cut branches eventually root and leaf out providing vegetative cover and secondary reinforcement as well.
VEGETATED GABION BASKETS 	Gabions are rectangular baskets made of triple-twisted hexagonal or welded-wire mesh of heavily galvanized and sometimes poly-vinyl chloride coated, steel wire. Gabions are delivered as flat wire panels that are folded into the basket form, filled with rock, and laced shut. Vegetation can be incorporated by placing cuttings through the mesh of the basket during filling or between the baskets after the baskets have been laced shut. These pervious structures can be used singly or stacked like building blocks.
LIVE CRIB WALL 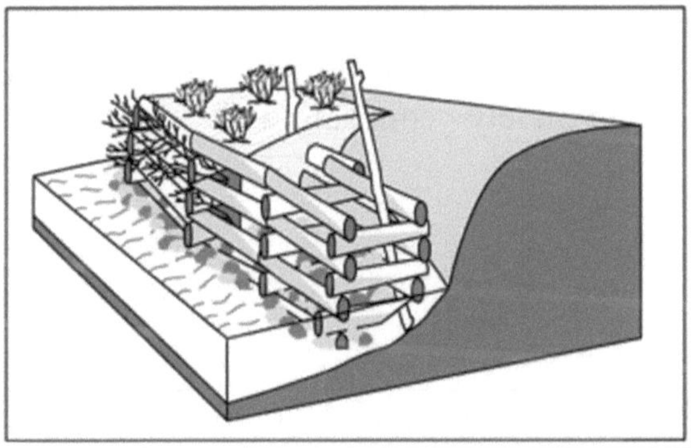	A crib wall is a gravity retaining structure consisting of a hollow, box-like interlocking arrangement of structural beams, e.g., logs. The interior of the crib-wall is filled with rock or soil. In *conventional* cribwalls, the structural members are fabricated from concrete, wood logs, and dimensioned timbers (usually treated wood). In *live* cribwalls, the structural members are usually untreated log or timber members. The structure is filled with a suitable backfill material, and live branch cuttings are inserted through openings between logs at the front of the structure and imbedded in the crib fill. These cuttings eventually root; the growing roots gradually permeate and reinforce the fill within the structure.
LIVE BRUSHLAYERING 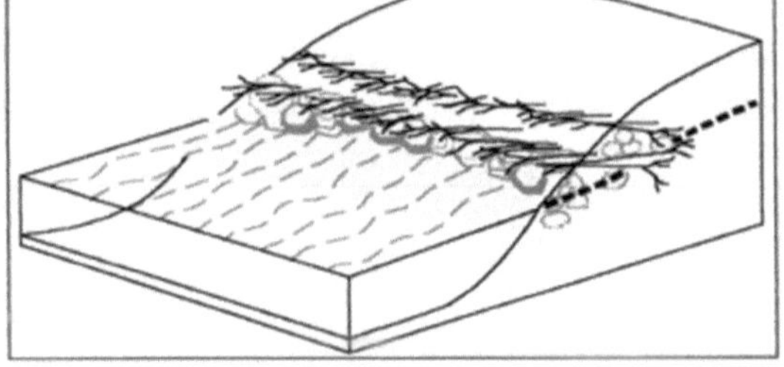	Live brushlayers are rows of live woody cuttings that are layered, alternating with successive lifts of soil fill, to construct a reinforced slope or embankment. The branches are placed perpendicular to the run of terraced lift in a criss-cross pattern so the tips protrude just beyond the face of the slope. The final cross section of the slope ranges to a maximum of 2H:1V. The live branch tips of the brush layer protrude in horizontal rows ranging from 0.5 m to 1.2 m apart, vertically. Vertical spacing depends on slope gradient and soil conditions.

WILLOW POSTS AND POLES 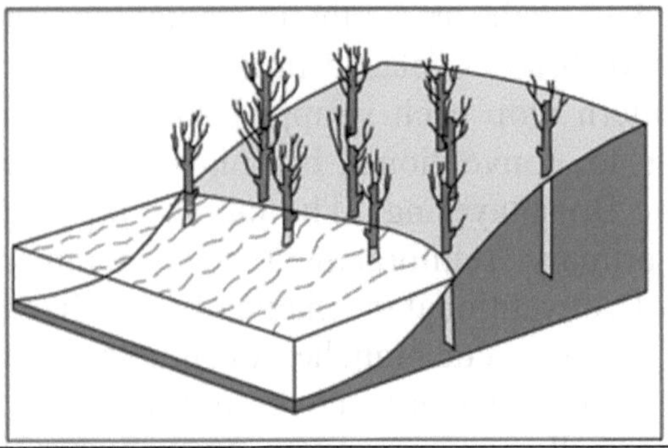	Willow (*Salix* spp.) trees and shrubs may be propagated by planting cuttings. Although smaller (<4 cm (1.5 in)) diameter cuttings (stakes) grow more vigorously than older, larger materials (posts and poles), larger materials provide better mechanical bank protection during the period of plant establishment. Dense arrays of posts or poles reduce velocities near the bank or bed surface, and long posts or poles reinforce banks against shallow mass failures or bank slumps.
LIVE SILTATION 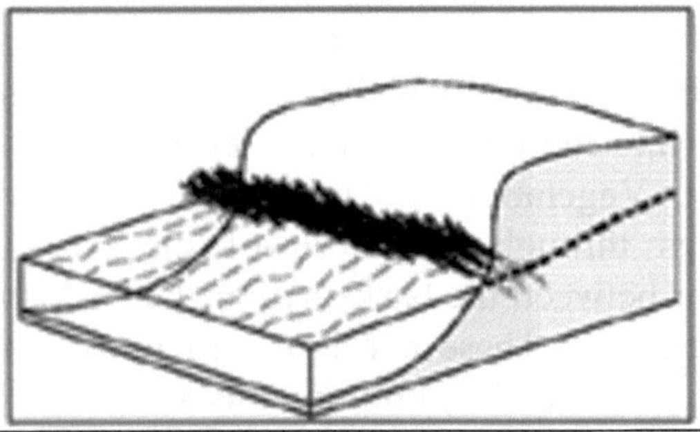	Live siltation is a revegetation technique used to secure the toe of a stream bank, trap sediments, and create fish rearing habitat. The system can be constructed as a living or a nonliving brushy system at the waters edge. The basal ends of cut branches are inserted into a trench excavated parallel to the toe of the stream bank which is then backfilled with earth. The brushy tips increase bank roughness, slow velocity, and promote sediment deposition.
TRENCH FILL REVETMENT 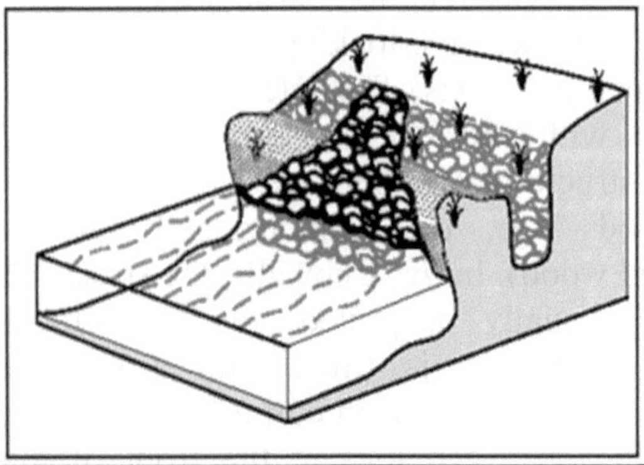	Trench fill revetments are constructed by excavating a trench along the top of the bank and placing stone riprap in the trench. As the bank erodes, the stone is undercut and "launches" down the bankline, resulting in a more gradual, protected slope. Earth removed for excavation of the trench may be used to cover the riprap, thus completely concealing it until it is launched.
VEGETATED FLOODWAYS 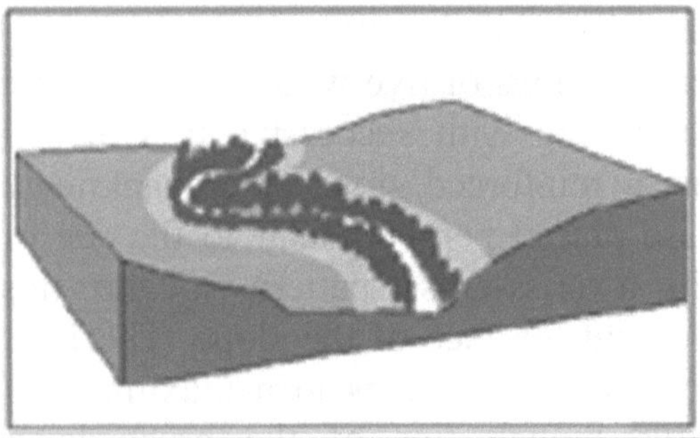	Confining floodwaters to a broad floodway bordered by levees or topographic highs is attractive because the portion of the floodway not normally inundated can support vegetation and thus provide wildlife habitat or recreational opportunities. Floodways may be created by constructing levees, floodwalls, or by excavation. Excavation consists of creating terraces or benches along an existing channel or a completely new flood channel (bypass). Roadway embankments sometimes serve a dual purpose by defining a floodway.

BANK ARMOR AND PROTECTION

VEGETATION ALONE

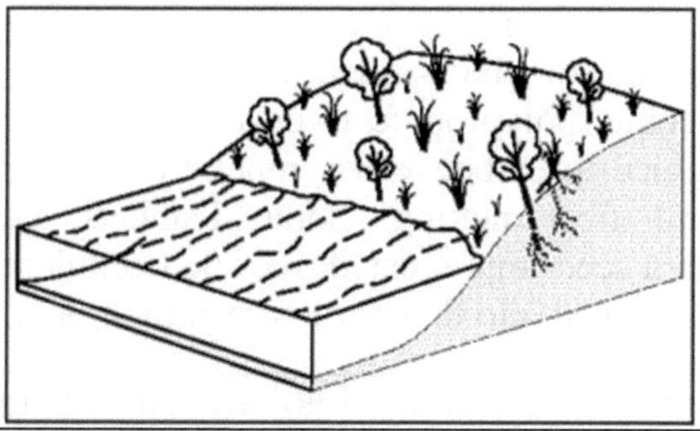

Vegetation can be viewed as a living, organic ground cover consisting of grasses/legumes, forbs, and/or woody plants. Vegetation is established on bare soils in order to help prevent surficial erosion, minimize shallow seated mass movement, provide habitat, and enhance esthetics or visual appearance. Vegetation can be used alone under special circumstances, but it also lends itself well to conjunctive use with other erosion control techniques in a mutually beneficial manner. Living plants can be used in conjunction with nearly every type of groundcover.

LIVE STAKING

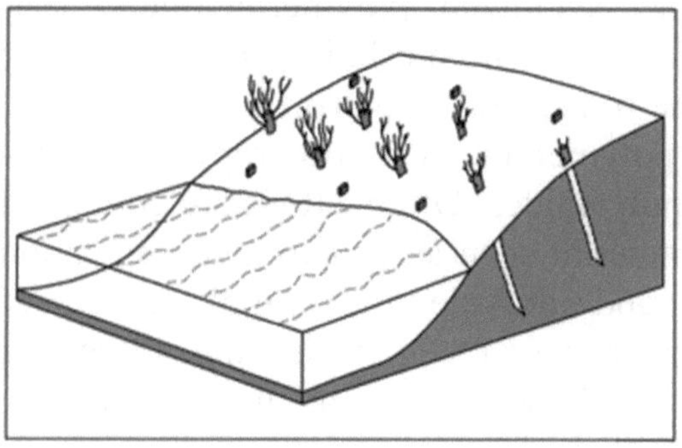

Live stakes are pieces of freshly cut woody plant stem planted in the ground or in erosion control structures, varying from about 30 to 76 cm long and typically 2.5–6.5 cm in diameter. They are inserted in such a way that a portion of the stem is buried, and a portion is exposed. The buried portion can take root, and the exposed portion can develop into a bushy riparian plant. Using a system of live stakes creates a root mat that stabilizes the soil by reinforcing and binding soil particles together. Roots can also aid stabilization by extracting excess soil moistureand by binding earthen fill to existing native soils.

LIVE FASCINES

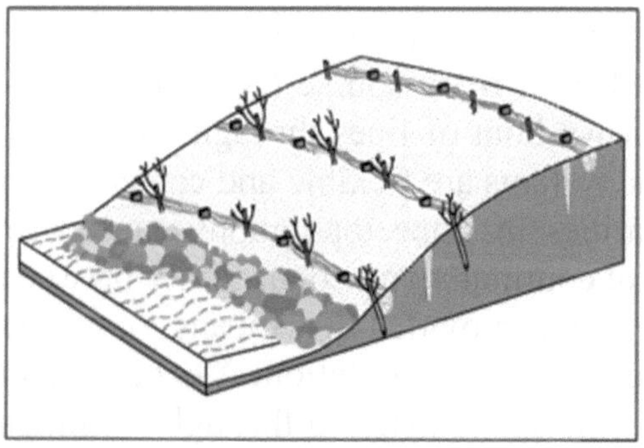

Live fascines are bundles of live branch cuttings placed in long rows in shallow trenches across the slope on contour or at an angle. Fascines are used for biotechnical stabilization of slopes and stream banks. Fascines are utilized as a resistive measure to protect the toe and face of an eroding stream bank; they are also very effective for erosion control on long bank slopes above annual high water. The terraces formed by the rows of fascines will trap sediment and detritus, promoting vegetative establishment. The partially buried bundles root and grow, providing strong, long-term protection.

LIVE BRUSH MATTRESS

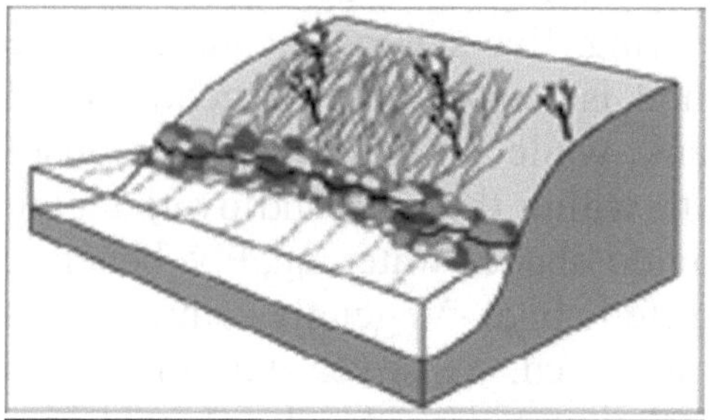

A live brush mattress, or brush mat, consists of a thick (6 in. to 1 foot) blanket of living brushy cuttings and soil fill placed on a stream bank or lake shore. This is a bank armoring technique for scour control and revegetation that is constructed using live willow branches or other species that easily root from cuttings. The dense layer of brush increases roughness, reducing velocities at the bank face, and protecting it from scouring, while trapping sediment and providing habitat directly along the waters' edge. Brush mattresses are used to simultaneously revegetate and armor the bank.

EROSION CONTROL BLANKETS 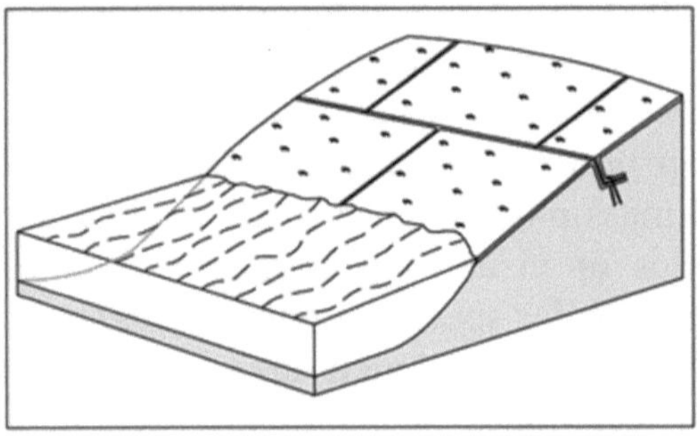	Erosion control blankets (ECBs) are a type of rolled erosion control product consisting of flexible nets or mats, manufactured from both natural and synthetic materials, which can be brought to a site, rolled out, and fastened down on a slope. ECBs are manufactured typically of fibers such as straw, wood, excelsior, coconut, polypropylene, or a combination, stitched, or glued to or between synthetic netting or woven natural fiber netting. Some blankets are available with a seed-impregnated, recycled cellulose medium incorporated into their structure at their base.
TURF REINFORCEMENT MATS 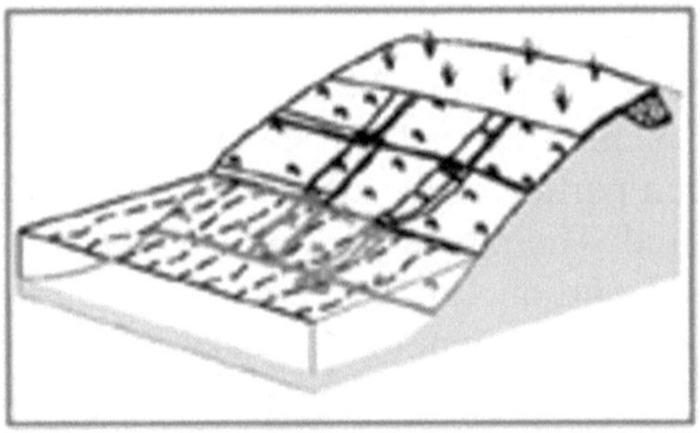	Turf Reinforcement Mats (TRMs) are similar to Erosion Control Blankets, but they usually are intended for lining channels. They are composed of ultraviolet (UV) stabilized polymeric fibers, filaments, nettings and/or wire mesh, integrating together to form a three-dimensional matrix 5–20 mm thick. The types of polymer include polypropylene, polyethylene, and polyamides. Often TRMs are combined with organic material such as coir to aide vegetation establishment and provide the initial temporary erosion control necessary to resist the forces of running water until the vegetation, usually grasses, can become established.
VEGETATED ACBs 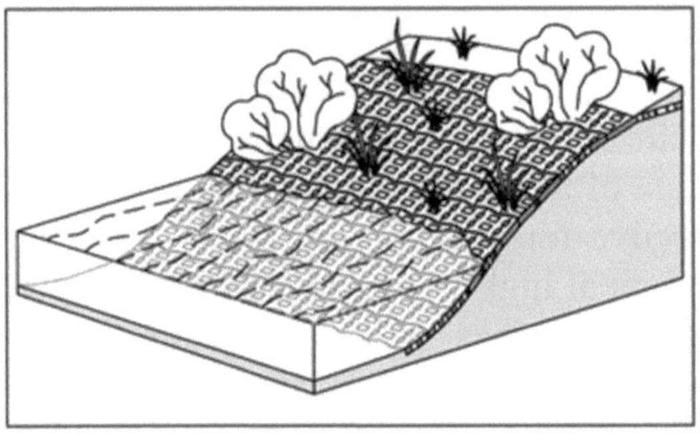	An articulated concrete block (ACB) system consists of durable concrete blocks that are placed together to form a matrix overlay or armor layer. The blocks abut or fit together to form a continuous blanket or mat. The blocks are placed on a filter course (typically a geofabric) to prevent washout of fines through the blocks. Articulated block systems are flexible and can conform to slight irregularities in slope topography caused by settlement. Lateral continuity and connectivity between individual blocks are provided by means of either cabling or some form of articulation. Vegetation in the form of live cuttings is inserted through openings in the blocks into the native soil beneath the blocks.
VEGETATED RIPRAP 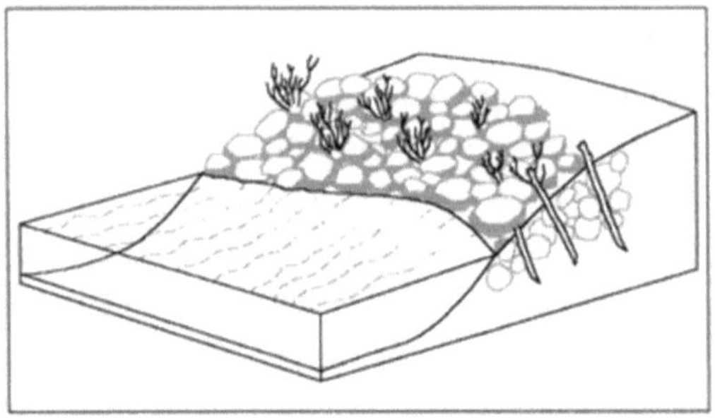	A layer of stone and/or boulder armoring that is vegetated, optimally during construction, using pole planting, brushlayering, live-staking techniques. The goal of this method is to increase the geotechnical stability of the bank, while simultaneously establishing riparian growth within the rock and overhanging the water, to provide shade, water quality benefits, and fish and wildlife habitat. Vegetative riprap combines the widely accepted, resistive, and continuous rock revetment techniques with deeply planted biotechnical techniques.

SOIL & GRASS COVERED RIPRAP 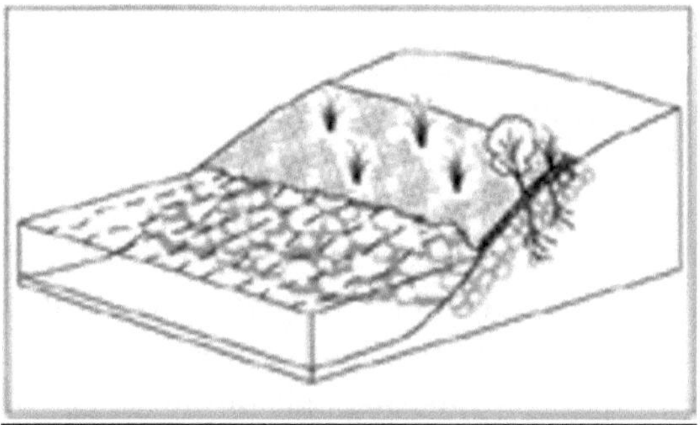	Two configurations have been used: (1), an ordinary riprap blanket is covered with a layer of soil 1–2 ft thick from the top of the revetment down to base flow elevation, or (2) a crown cap of soil and plant material is placed over a riprap toe running along the base of a steep bank, effectively reducing bank angle. Soils used for fill should not be highly erosive. A variety of methods may be used to establish plant materials including seeding and mulching, sodding, and incorporation of willow cuttings or root stock in the fill materials.
COBBLE OR GRAVEL ARMOR 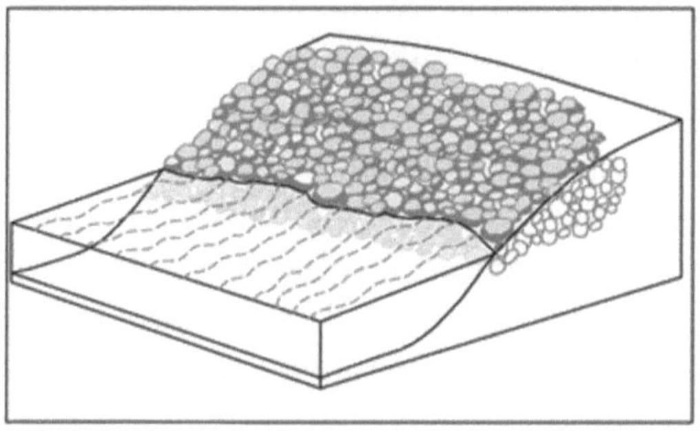	Riprap stone blanket or riprap revetment is the most widely used form of stream bank protection. Although properly designed riprap revetments can be quite effective, the color and angularity of quarried stone is not a natural component of stream corridors. Cobbles are natural stones larger than 64 mm (2.5 in.) in diameter that have been rounded by the abrasive action of flowing water, while gravel is material smaller than cobble but larger than sand (larger than about 5 mm) (0.2 in). Rounded river cobble or gravel blanket presents a more natural appearance and can be nearly as effective as angular stone riprap revetment.
VEGETATED GABION MATTRESS 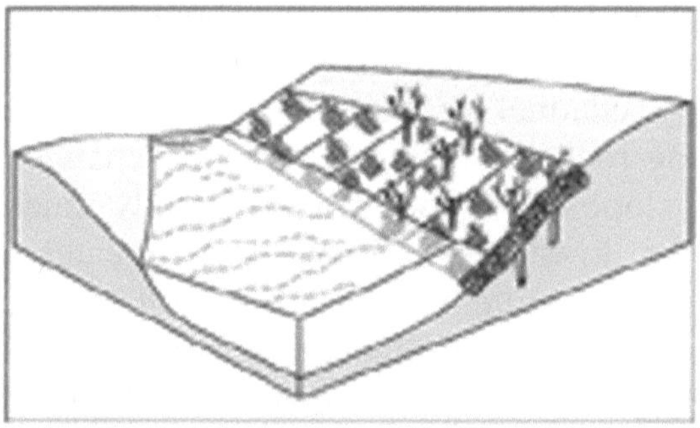	Gabion mattresses are designed to protect the bed or lower banks of a stream against erosion. They differ from gabion baskets, which are meant to protect the mid to upper bank. Gabion mattresses are also shallower (0.5–1.5m deep) rectangular containers made of welded wire mesh and filled with rock. A gabion mattress can be used as either a revetment to stabilize a stream bank, or when used in a channel, to decrease the effects of scour. Live cuttings are introduced through the rock-filled mattress and inserted into native soil beneath.
ROOTWAD REVETMENT 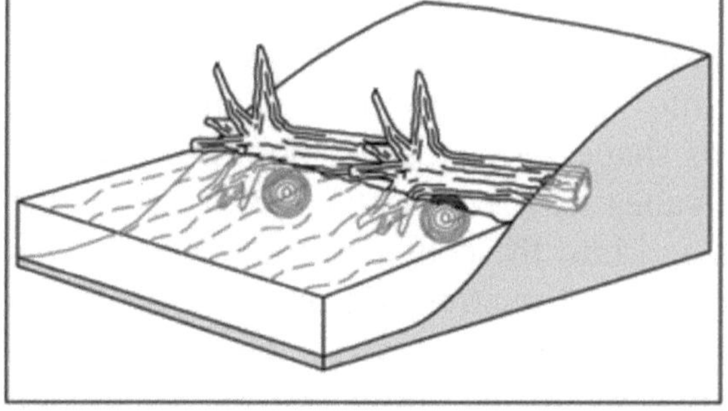	Rootwad revetments are structures constructed from interlocking tree materials. They are primarily intended to resist erosive flows and are usually used on the outer bends of streams. Rootwad revetments are a root ball, commonly called a rootwad, of a tree plus a portion of the trunk in combination with rocks and biotechnical methods to construct continuous bank protection. Rootwad revetments are designed such that the rootwads are oriented upstream into the stream flow and are frequently placed against each other for continuous armoring along the entire length of a meander bend.

RIPARIAN BUFFER AND STREAM CORRIDOR OPPORTUNITIES	
LIVE GULLY FILL REPAIR 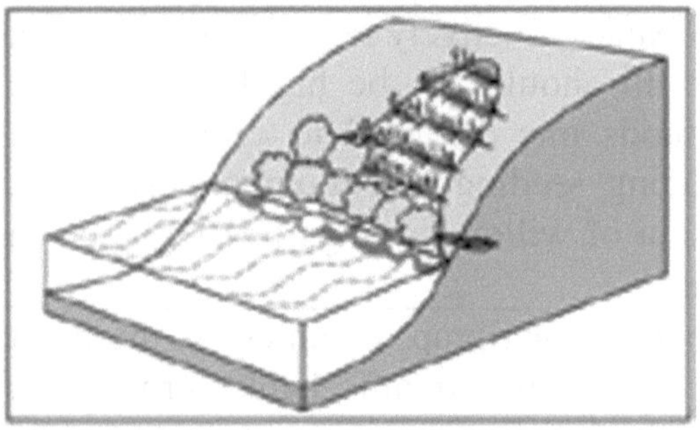	Live Gully Fill Repair consists of alternating layers of live branch cuttings and compacted soil. This reinforced fill can be used to repair small gullies. The method is similar to branch packing (a method for filling small holes and depressions in a slope) but is more suitable for filling and repairing elongated voids in a slope such as gullies. Gully treatment must include correcting or eliminating the initial cause of the gully as well as the gully itself. Gullies are likely to have side gullies that require treatment; but priority should be given to the main channel.
BOULDER CLUSTERS 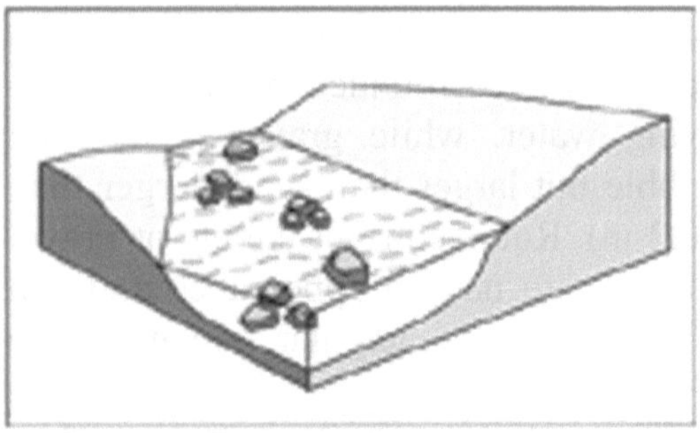	Large boulders may be placed in various patterned clusters within the base flow channel of perennial stream. Natural streams with beds coarser than gravel often feature large roughness elements like boulders that provide hiding cover and velocity shelters for fish and other aquatic organisms. If a constructed or modified channel lacks such features, adding boulder clusters may be an effective and simple way to improve aquatic habitat.
NEWBURY ROCK RIFFLES 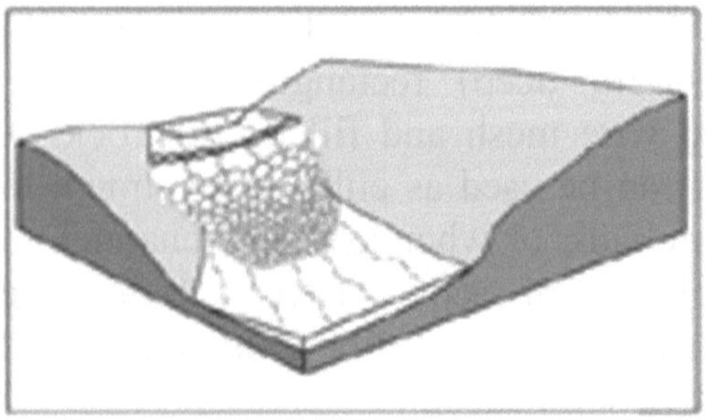	Newbury rock riffles are ramps or low weirs with long aprons made from riprap or small boulders that are constructed at intervals approaching natural riffle spacing (5 to 7 channel widths). The structures are built by placing rock fill within an existing channel. The upstream slope of the rock fill is typically much steeper than the downstream slope, which creates a longitudinal profile quite similar to natural riffles. These structures provide limited grade control, pool and riffle habitat, and visual diversity in otherwise uniform channels.
CROSS VANES 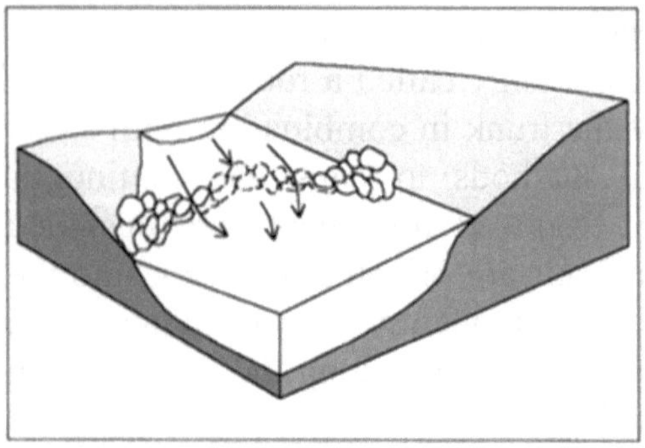	Cross vanes are "V" shaped rock structures stretching across the width of the stream, with the lowest part of the structure being at the base of the "V" at the point farthest upstream. The downstream ends of the "V" are keyed into the stream banks at bank-full elevation. This shape forms a scour pool inside of the "V." Cross vanes redirect water away from the stream banks and into the center of the channel. This serves to decrease shear stress on unstable banks, as well as creating aquatic habitat in the scour pools formed by the redirected flow. Cross vanes are particularly useful for modifying flow patterns upstream from bridges.

SLOPE STABILIZATION

SLOPE FLATTENING

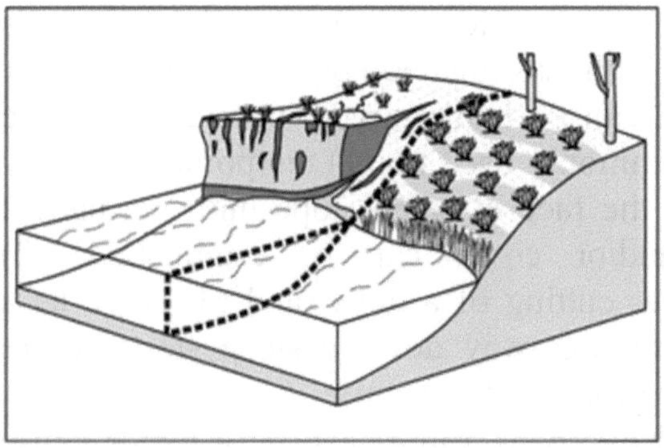

Flattening or bank reshaping stabilizes an eroding stream bank by reducing its slope angle or gradient. Slope flattening is usually done in conjunction with other bank protection treatments, including installation of toe protection, placement of bank armor, revegetation or erosion control, and/or installation of drainage measures. Flattening or gradient reduction can be accomplished in several ways: (1) by removal of material near the crest, (2) by adding soil or fill at the bottom, or (3) by placing a toe structure at the bottom and adding a sloping fill behind it. Right-of-way constraints may limit or pre-clude the first two alternatives because both entail either moving the crest back or extending the toe forward.

DIVERSION DIKE

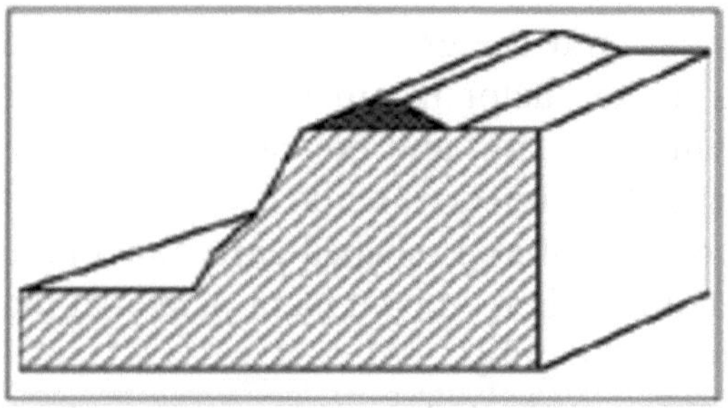

A diversion dike is a low berm (or ditch and berm combination) that is constructed along the crest or top of a stream bank. The purpose of a diversion is to intercept and divert runoff away from the face of a steep slope or stream bank. Diversion dikes are constructed from compacted earthen fill and should be used on drainage areas of 2 ha or less. In addition to protecting the face of a stream bank from overbank runoff, diversions may also improve general slope stability by preventing runoff from infiltrating into and saturating the face of the bank.

TRENCH DRAIN

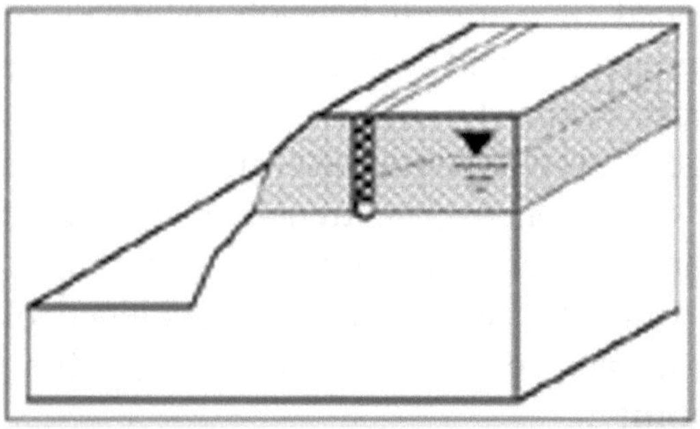

A drainage trench is excavated parallel to and just behind the crest of a stream bank. Ideally, the bottom of the trench should be keyed into an impermeable layer in the slope. The trench should be backfilled with a coarse graded aggregate that meets filtration criteria; i.e., it should allow unimpeded flow of groundwater while excluding fines. Alternatively, the trench can first be lined with a filter fabric that meets the filtration require-ments and then be backfilled with a coarse aggregate. The purpose of the trench is to intercept and divert shallow seepage away from the face of the stream bank.

CHIMNEY DRAIN

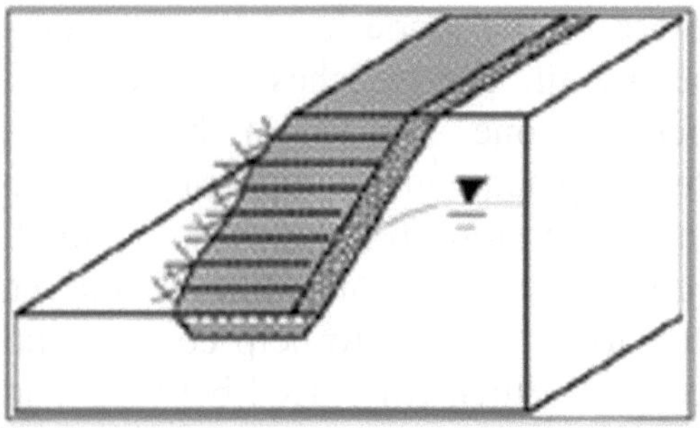

A chimney drain is a subsurface drainage course placed between a natural slope (or stream bank) and an earthen buttress fill or other retaining structure (e.g., log crib wall). A drainage blanket, sloped sheet drain, and strip drain are types of subsurface drainage courses. Typically a chimney drain is a near vertical drain that feeds into a collection system at its base whereas a sloped sheet drain is inclined back at an angle. A sub-surface drain may be continuous across the slope, or it may consist of discontinuous drainage strips that are placed against the natural slope at periodic intervals.

SLOPE DRAIN

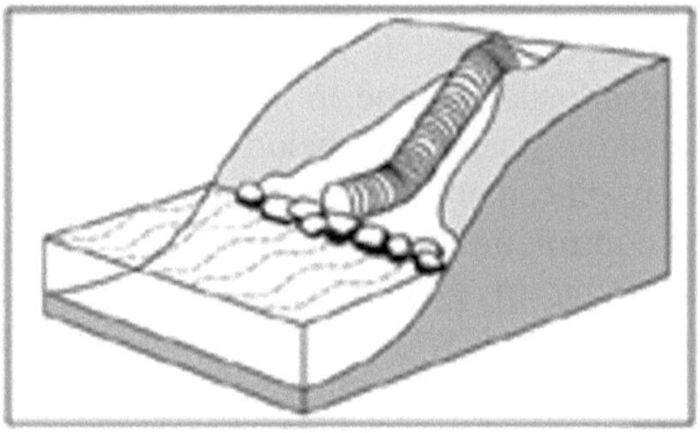

A slope drain is a drainage system used to collect and transport storm runoff down the face of a slope. This system usually consists of a berm at the top of the slope (or stream bank) and a flexible pipe with end sections and outlet protection. A pipe slope drain is constructed with corrugated pipes (polymeric or metallic) and can be temporary or permanent. Slope drains are commonly used to: (1) temporarily convey runoff down the face of steep slope until permanent protection and/or cover can be established, (2) prevent further cutting of a gully, and (3) serve as a permanent drainage-way down a steep slope where visual appearance is not a factor.

DROP INLET

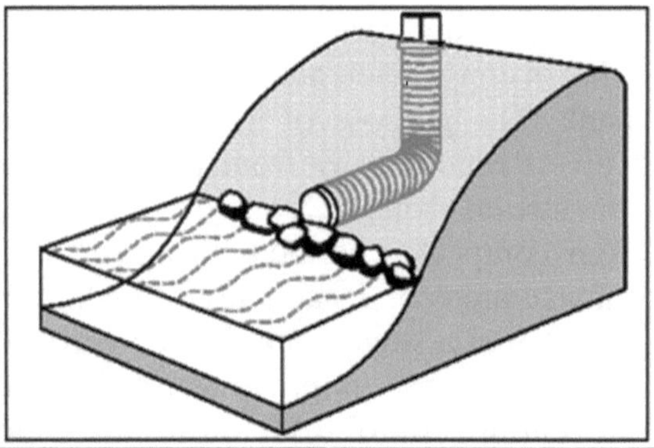

Concentrated overbank runoff can be a major cause of erosion, especially along deeply incised channels. Runoff passing over the top of banks frequently triggers gully development and expansion, and water that is ponded at the top of high, steep banks and infiltrates or seeps into the ground behind the slope face is often a major factor in erosion by piping or slope failure. Gully erosion and downcutting can be addressed using a drop inlet, which is a water control structure that consists of an L-shaped corrugated pipe passing through an earthen embankment placed at the downstream end of the gully.

FASCINES WITH SUB DRAIN

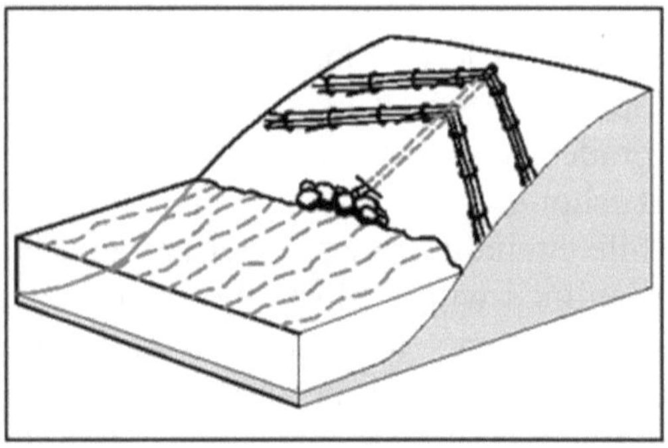

Rows of fascines are installed on contour on a slope in the conventional manner. In addition, a subsurface drain, oriented downslope and perpendicular to the fascines, is placed in a trench beneath the rows of fascines to intercept and collect seepage. The subsurface drain consists of a perforated pipe wrapped in a geo-composite drainage medium placed at the bottom of a trench. The trench is backfilled with clean, coarse aggregate or gravel. This system is used on very wet sites where there is evidence of substantial subsurface seepage that is causing piping and slope instability.

STONE-FILL TRENCHES

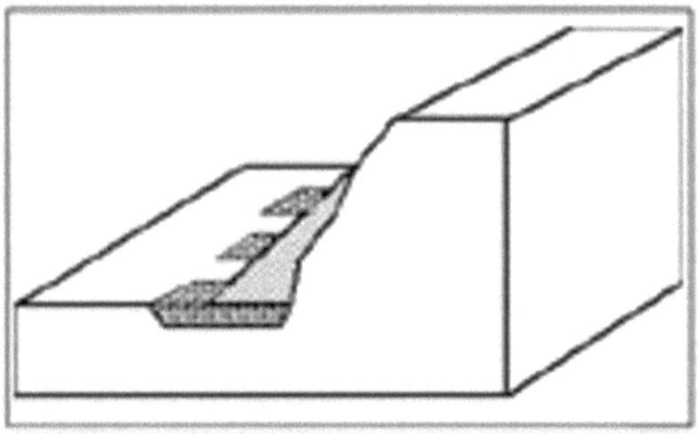

Stone-fill trenches are rock-filled trenches placed at the base of a stream bank, usually within a failed section of the toe. A series of trenches are excavated at or within the toe of the slope in a direction perpendicular to the stream. The trenches are backfilled with crushed rock or stone. The toe of the slope is then reconstructed by placing and compacting earthen fill within and atop the stone-fill trenches. A small, longitudinal riverside plug or stone dike should be used between the stone trenches to help contain and protect the toe of the earthen fill placed between and atop the stone trenches.

ADDENDUM

FIBER ROLLS

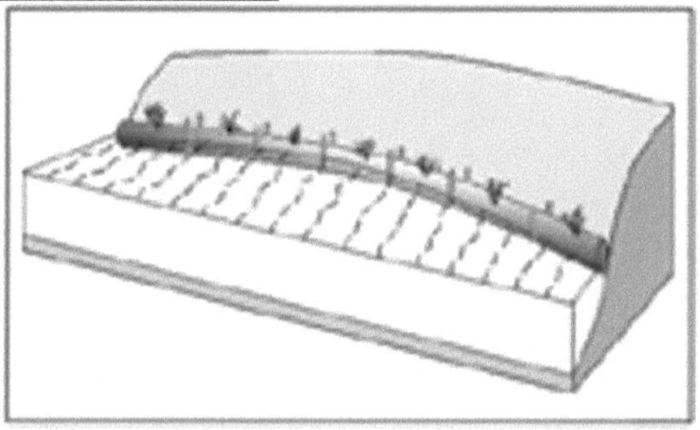

Fiber rolls are manufactured, elongated cylindrical structures that are placed at the toe of a stream bank. They are typically made of coconut husk fibers bound together with a high strength synthetic netting. When natural coir mesh is used for the exterior, they are usually referred to as coir rolls. The rolls are staked, cabled, wedged, or otherwise anchored in place at the toe.

POLE DRAINS

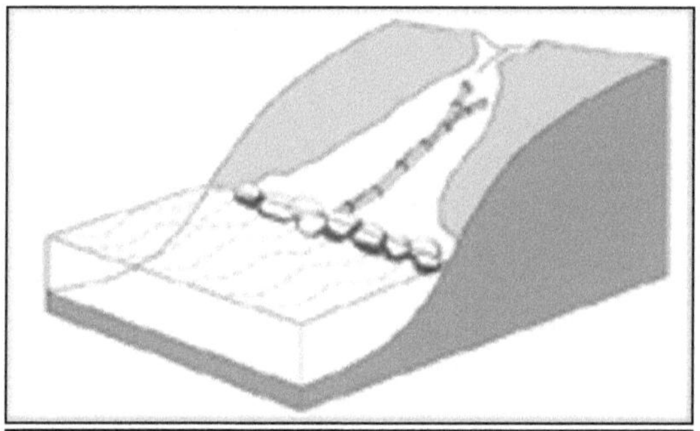

Live pole drains collect subsurface drainage and concentrated surface flow and channel it to the base of a stream bank or slope. Because they are long and fibrous, the bundles act like a conduit. The water will take the path of least resistance, in this case, running down the bundle. The roots in willow bundles act like a filter fabric, stabilizing fine particles and reducing piping and sapping.

GEOCELLULAR CONTAINMENT

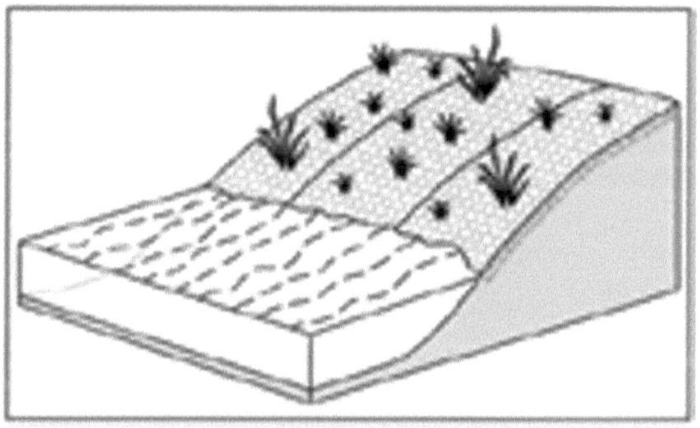

Geocellular confinement systems (GCS) are a permanent erosion control practice for stabilizing slopes as steep as 0.5V:1H. They can be filled with rock, gravel, topsoil, or a combination of materials to form an erosion resistant surface. The cells can be formed by a lattice or grid of wood beams or alternatively by a 3-dimensional and expandable polymer grid system that is staked in place on the slope.

MEANDER RESTORATION

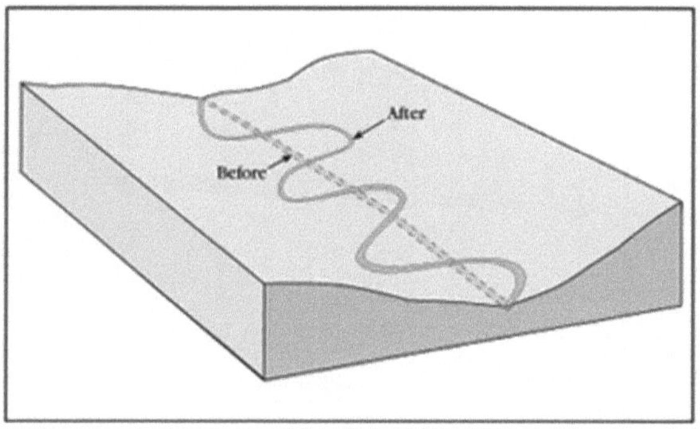

Meanders are broad, looping (sinuous) bends in a stream channel. Meandering is a form of channel slope adjustment with more sinuous channel paths leading to decreased gradient. Fluvial and ecological functions are integrally related to the highly diverse spatial and temporal patterns of depth, velocity, bed material, and cover found in meanders. Meander restoration consists of reconstructing meandering channels that have typically been straightened by humans.

APPENDIX 2 – ALTERNATIVE NAMES

SUMMARY OF ENVIRONMENTALLY SENSITIVE HILLSIDE AND STREAM BANK PROTECTION MEASURES
(adapted from NCHRP, 2005)

ALTERNATIVE NAMES AND DESIGNATIONS (AKAs)	
COCONUT FIBER ROLLS	**Coir Logs**
EROSION CONTROL BLANKETS and **TURF REINFORCEMENT MATS**	**Rolled Erosion Control Products (RECPs)**
LARGE WOODY DEBRIS STRUCTURES	**Large Wood Structures**
SPUR DIKES	**Boulder Spurs/Bank Spurs**
VEGETATION ALONE	**Revegetation**
WILLOW POSTS AND POLES	**Live Pole Planting**

W. Goldsmith et al., *Bioengineering Case Studies: Sustainable Stream Bank and Slope Stabilization*,
DOI 10.1007/978-1-4614-7996-3, © Springer Science+Business Media New York 2014

APPENDIX 3 – SUPPLEMENTARY GLOSSARY OF NEW TERMS AND TREATMENTS

(Not included in APPENDIX 1 – NCHRP Rept. No. 544 ESCBMs)

Bioswales

A multipurpose, elongated shallow ditch that collects, stores, filters, cleanses, and infiltrates water as runoff from the built environment. Vegetation stabilizes the soil, provides nutrient uptake, maintains infiltration properties, and serves habitat value.

Blueberry Sod

Low-bush blueberry and other related native species that are commercially available when blueberry farms are converted to development. They are pegged in place on steep and bare banks to stabilize shaded areas and to match the local surrounding understory.

Boulder Cascades [similar to NEWBURY ROCK RIFFLES]

Placement of boulder and rock aprons immediately downstream of check dams that mimic natural cascades and that dissipate energy of flowing water

Channel Profile Modification

Modification of the channel profile in a cross-channel direction to improve stream function. The channel width can be widened along its bottom to accommodate increased flow; the sides can also be graded back to a more stable angle.

Coir Netting [similar to EROSION CONTROL BLANKET]

A ground net manufactured with coconut fibers that is used to hold long-straw mulch and grass seeds in place on a slope.

Drainage Swales

Construction of land forms to maintain stable runoff patterns and support original plant distributions. Includes use of rock lined, curvilinear ditches with porous bottoms to allow both runoff and infiltration to sustain plant life.

Fiberschines, Straw Wattles [similar to COCONUT FIBER ROLLS]

Coir (coconut) fiber, rice straw, or some other inert and relatively strong and stable fiber is used instead of dead branches or live woody cuttings.

Gabion Check Dam

A check dam constructed of rock filled, gabion baskets, designed to provide grade control, arrest channel down-cutting, and protect the toe of adjacent channel side slopes.

W. Goldsmith et al., *Bioengineering Case Studies: Sustainable Stream Bank and Slope Stabilization*,
DOI 10.1007/978-1-4614-7996-3, © Springer Science+Business Media New York 2014

Green Roof

A lightweight engineered growing substrate atop a building that supports alpine type vegetation and allows the building to absorb rainfall and slowly release moisture via evapo-transpiration.

Landform Grading

Geomorphic modification or regrading to eliminate erosion scars and to mimic stable natural landforms that are compatible with local slope hydrogeology.

Live Smiles

Woven willow fences, usually less than 10 ft long, formed into a "smile" (catenary curve) that are used to hold slumping soils on a slope. In addition to the structural stability these measures provide, they also have the advantage of growing and sending down roots which reinforce the soil and stabilize a slope.

Modified VMSE (soil flaps)

A geogrid or net is placed on a slope surface and then pulled down tightly against the surface by means of driven line anchors. This anchoring procedure develops tension in the overlying geogrid which in turn imparts a confining stress on the ground surface. Soil flaps can be used in lieu of the conventional, horizontal placement of geogrid reinforcements between successive lifts of soil.

Riparian Plantings [similar to VEGETATION ALONE]

Riparian plants that provide nutrient uptake, contaminant phytoremediation, targeted habitat functions, floodway stabilization, and additional bank erosion protection along waterways and channels.

Rocked, Stepped-Pools

A series of low-head, boulder drop-structures connected by a low-gradient channel.

Rock Ramp [similar to NEWBURY ROCK RIFFLES]

An inclined rock ramp or apron that dissipates energy of flowing water and provides limited grade control.

Sinuous Channel

An artificial channel with a sinuous (curved) planform designed to lengthen the flow path, decrease the flow gradient, and to convey concentrated flow at a noneroding velocity. The channel can be lined with a coir geotextile and/or coarse gravel to increase resistance to tractive stresses.

Soil Amendments

Addition of compost to a soil that increases infiltration rates, rooting depth, fertility, and water holding capacity. These factors are all critical to revegetation on adverse sites, especially slopes and exposed subsoils. Compost additions encourage deep rooting of native grasses and willows.

Stream Realignment

Excavation of a low flow channel to divert a stream back into its historic location. The historic dimensions, e.g., bank-full width, slope, and sinuosity, as determined from a "reference reach," are used to guide design and implementation. The inner bend(s) of the realigned reach can be excavated to allow floodwaters to access the floodplain.

Index

W. Goldsmith et al., *Bioengineering Case Studies: Sustainable Stream Bank and Slope Stabilization*,
DOI 10.1007/978-1-4614-7996-3, © Springer Science+Business Media New York 2014

MIX
Papier aus verantwortungsvollen Quellen
Paper from responsible sources
FSC® C105338

If you have any concerns about our products,
you can contact us on
ProductSafety@springernature.com

In case Publisher is established outside the EU,
the EU authorized representative is:
Springer Nature Customer Service Center GmbH
Europaplatz 3, 69115 Heidelberg, Germany

Printed by Libri Plureos GmbH
in Hamburg, Germany